Ordered Algebraic Structures

Mathematics and Its Applications

Volume 55

Ordered Algebraic Structures

Proceedings of the
Caribbean Mathematics Foundation Conference
on Ordered Algebraic Structures,
Curaçao, August 1988

edited by

Jorge Martinez

Department of Mathematics, University of Florida,
Gainesville, U.S.A.

KLUWER ACADEMIC PUBLISHERS
DORDRECHT / BOSTON / LONDON

Library of Congress Cataloging in Publication Data

Caribbean Mathematics Foundation Conference on Ordered Algebraic Structures (1988 : Princess Beach Hotel, Curaçao)
Ordered algebraic structures : proceedings of the Caribbean Mathematics Foundation Conference on Ordered Algebraic Structures, Curaçao, August 1988 / edited by Jorge Martinez.
p. cm. -- (Mathematics and its applications)
ISBN 0-7923-0489-6 (U.S.)
1. Ordered algebraic structures--Congresses. I. Martinez, Jorge, 1945- . II. Title. III. Series: Mathematics and its applications (Kluwer Academic Publishers)
QA172.C37 1988
512'.2--dc20 89-24396

ISBN 0-7923-0489-6

Published by Kluwer Academic Publishers,
P.O. Box 17, 3300 AA Dordrecht, The Netherlands.

Kluwer Academic Publishers incorporates
the publishing programmes of
D. Reidel, Martinus Nijhoff, Dr W. Junk and MTP Press.

Sold and distributed in the U.S.A. and Canada
by Kluwer Academic Publishers,
101 Philip Drive, Norwell, MA 02061, U.S.A.

In all other countries, sold and distributed
by Kluwer Academic Publishers Group,
P.O. Box 322, 3300 AH Dordrecht, The Netherlands.

Printed on acid-free paper

Printed in The Netherlands

This volume is dedicated to the memory of my father,

who would have enjoyed this event more than anyone else.

SERIES EDITOR'S PREFACE

'Et moi, ..., si j'avait su comment en revenir, je n'y serais point allé.'

Jules Verne

The series is divergent; therefore we may be able to do something with it.

O. Heaviside

One service mathematics has rendered the human race. It has put common sense back where it belongs, on the topmost shelf next to the dusty canister labelled 'discarded non-sense'.

Eric T. Bell

Mathematics is a tool for thought. A highly necessary tool in a world where both feedback and non-linearities abound. Similarly, all kinds of parts of mathematics serve as tools for other parts and for other sciences.

Applying a simple rewriting rule to the quote on the right above one finds such statements as: 'One service topology has rendered mathematical physics ...'; 'One service logic has rendered computer science ...'; 'One service category theory has rendered mathematics ...'. All arguably true. And all statements obtainable this way form part of the raison d'être of this series.

This series, *Mathematics and Its Applications*, started in 1977. Now that over one hundred volumes have appeared it seems opportune to reexamine its scope. At the time I wrote

> "Growing specialization and diversification have brought a host of monographs and textbooks on increasingly specialized topics. However, the 'tree' of knowledge of mathematics and related fields does not grow only by putting forth new branches. It also happens, quite often in fact, that branches which were thought to be completely disparate are suddenly seen to be related. Further, the kind and level of sophistication of mathematics applied in various sciences has changed drastically in recent years: measure theory is used (non-trivially) in regional and theoretical economics; algebraic geometry interacts with physics; the Minkowsky lemma, coding theory and the structure of water meet one another in packing and covering theory; quantum fields, crystal defects and mathematical programming profit from homotopy theory; Lie algebras are relevant to filtering; and prediction and electrical engineering can use Stein spaces. And in addition to this there are such new emerging subdisciplines as 'experimental mathematics', 'CFD', 'completely integrable systems', 'chaos, synergetics and large-scale order', which are almost impossible to fit into the existing classification schemes. They draw upon widely different sections of mathematics."

By and large, all this still applies today. It is still true that at first sight mathematics seems rather fragmented and that to find, see, and exploit the deeper underlying interrelations more effort is needed and so are books that can help mathematicians and scientists do so. Accordingly MIA will continue to try to make such books available.

If anything, the description I gave in 1977 is now an understatement. To the examples of interaction areas one should add string theory where Riemann surfaces, algebraic geometry, modular functions, knots, quantum field theory, Kac-Moody algebras, monstrous moonshine (and more) all come together. And to the examples of things which can be usefully applied let me add the topic 'finite geometry'; a combination of words which sounds like it might not even exist, let alone be applicable. And yet it is being applied: to statistics via designs, to radar/sonar detection arrays (via finite projective planes), and to bus connections of VLSI chips (via difference sets). There seems to be no part of (so-called pure) mathematics that is not in immediate danger of being applied. And, accordingly, the applied mathematician needs to be aware of much more. Besides analysis and numerics, the traditional workhorses, he may need all kinds of combinatorics, algebra, probability, and so on.

In addition, the applied scientist needs to cope increasingly with the nonlinear world and the extra mathematical sophistication that this requires. For that is where the rewards are. Linear models are honest and a bit sad and depressing: proportional efforts and results. It is in the non-linear world that infinitesimal inputs may result in macroscopic outputs (or vice versa). To appreciate what I am hinting at: if electronics were linear we would have no fun with transistors and computers; we would have no TV; in fact you would not be reading these lines.

There is also no safety in ignoring such outlandish things as nonstandard analysis, superspace and anticommuting integration, *p*-adic and ultrametric space. All three have applications in both electrical engineering and physics. Once, complex numbers were equally outlandish, but they frequently proved the shortest path between 'real' results. Similarly, the first two topics named have already provided a number of 'wormhole' paths. There is no telling where all this is leading - fortunately.

Thus the original scope of the series, which for various (sound) reasons now comprises five sub-series: white (Japan), yellow (China), red (USSR), blue (Eastern Europe), and green (everything else), still applies. It has been enlarged a bit to include books treating of the tools from one subdiscipline which are used in others. Thus the series still aims at books dealing with:

- a central concept which plays an important role in several different mathematical and/or scientific specialization areas;
- new applications of the results and ideas from one area of scientific endeavour into another;
- influences which the results, problems and concepts of one field of enquiry have, and have had, on the development of another.

Now that a great deal of mathematics has been developed for all kinds of 'single' structures, the time seems to have come to pay a good deal of attention to the case of 'composite' structures; i.e. sets equipped with two (or more) different structures which are compatible in a suitable way. In the present case an order structure together with an algebraic one, usually a group or ring.

In the preface of a greatly related volume in this series (A.M.W. Glass and W.Ch. Holland (Eds.), Lattice-ordered Groups, KAP, 1989) I made a number of remarks on 'composite' structures in algebra and noted that, though in most cases exploration of the consequences of compatibility has only just begun, the case of lattice-ordered algebraic structures has reached a certain plateau of sophistication and maturity. That does not imply in any way that nothing important is happening, as the present result of a conference in Curaçao amply testifies.

One aspect of 'composite' structures seems to be that the compatibility tends to force all of the components to be more regular, beautiful, symmetric. In these terms the question has, it seems to me, hardly been explored at all. However, in the case that one of the components is order, it has certainly become clear, at the Curaçao conference for instance, that the phenomena become very different from what they would have been without such a relation imposing order.

The shortest path between two truths in the real domain passes through the complex domain.

J. Hadamard

La physique ne nous donne pas seulement l'occasion de résoudre des problèmes ... elle nous fait pressentir la solution.

H. Poincaré

Never lend books, for no one ever returns them; the only books I have in my library are books that other folk have lent me.

Anatole France

The function of an expert is not to be more right than other people, but to be wrong for more sophisticated reasons.

David Butler

Bussum, September 1989 Michiel Hazewinkel

CONTENTS

CHAPTER TWO: RINGS

ORDER IN CURAÇAO: A PREFACE

On the eighth of August, 1988, nearly three dozen mathematicians gathered in Curaçao, in the Netherlands Antilles, for a conference on Ordered Algebraic Structures. For five consecutive days 26 speakers presented papers on various aspects of that brand of mathematics which commonly falls under the label that gave its title to the conference.

During these lectures a number of mathematical ideas and topics were discussed; lattices, groups, rings of continuous functions, totally ordered groups, lattice-ordered groups, f-rings, ordered fields, orderability of groups, groups of order-preserving permutations, model-theoretic aspects of totally ordered groups, amalgamations of ordered groups, structure topologies on ordered groups, Riesz spaces and the theory of positive linear operators. The preceding list should not be taken as complete; it is haphazard in its assemblage, by design. There were mathematicians present who are frequently considered analysts; otheres who are categorized as topologists; still another group commonly thought of as algebraists. A large number of the participants would, I should like to think, resist being labelled by any of the three traditional disciplines of mathematics.

There were five problem sessions during the conference, which were reasonably lively, and during which no evidence surfaced that the participants were handicapped by the differences in their specialties or preferences. Instead, these individuals extended themselves beyond the particularity and slant of their training to converse intelligently with their collegues.

All of which suggest – in addition to the obvious fact that the conference had brought together some outstanding men and women – that a number of assumptions commonly made about Ordered Algebra are wrong, chief among them the belief that the "field" is not going anywhere. Secondly, I should think that the label itself, that umbrella under which we gather to perform as mathematicians, is misleading, and not very informative in any event. An individual no more deserves to be called an Ordered Algebrist than an Ordered Analyst or an Ordered Topologist. I therefore propose that the label be mothballed and go the way of all obsolete or meaningless labels. Perhaps all labels are absurd in mathematics today, but my subject is not labels, it is the conference which took place a year ago in Curaçao. There, a number of mathematicians came together and discussed some rather remarkable ideas, and the fabric of those ideas seemed to be woven by a common thread: the structural incorporation of an ordering, which made the techniques to be applied and, frequently, the results as well, quite unlike what they would have been without that ordering.

Some extraordinary people were in attendance. And, whereas I am opposed to rankings of men and women, I am as sentimental as the next fellow, and therefore have little trouble recognizing those who have influenced me. My wager is that most people gathered in Curaçao will share my sentiments. I celebrate the achievements of Conrad, Henriksen, Holland, Luxemburg & Ribenboim; the accomplishments of their students, colleages and co-authors speak volumes about the scale of their contributions to mathematics.

This volume contains twenty-one articles, and a large number of them are expository presentations. In a very important respect this is a matter of design. The conference was, after all, intended to be appreciated by a broad audience of mathematicians; although we were less successful than expected in drawing mathematicians from the Caribbean region and from Latin America, the goal of reaching a broad audience is thereby hardly discredited, and for this reason the proceedings of the conference were shaped to attempt to achieve such a goal.

The topics addressed during the conference and in this book as well, are all being actively investigated by their expositors. The observation, that this collection of articles mirrors the state of the art in ordered mathematics, is perhaps redundant; it is nonetheless important, as a political statement, to give evidence that a protean discipline such as ours is alive and well. I submit this text in evidence of that state. It will arrive on the market on the heels of two publications dealing with these subjects: the book by Anderson and Feil, Lattice-Ordered Groups: An Introduction, published by Kluwer, and the proceedings of the 1985 Bowling Green Workshop, Lattice-Ordered Groups: Advances and Techniques, edited by Glass and Holland, also a Kluwer publication. (The latter contains a bibliography with some 800 references!) These proceedings will probably appear simultaneously with yet another reference text on lattice-ordered groups, written by Darnel. All in all this activity gives substantial evidence, I should think, that the area is not only alive in research terms, but that its proponents are taking care to promote it widely and thoughtfully.

The papers of Anderson, Henriksen, Luxemburg and Powell in this volume are drawn from expository lectures given at the conference. A number of the articles are distillates of material which has either appeared or else will appear in full glory elsewhere. Others, like the contributions by Ball & Hager, Darnel and Redfield, are expository accounts of material which appears dispersed over a plurality of papers. The contributions by Ball & McCleary, Bernau & Huijsmans, Bradley & Prestel, Dauns, Glass, Gluschankof, Madden, McCleary, Rhemtulla, Schwartz and Wojciechowski are of original work, and they incorporate a generous supply of proofs. The papers by Dauns – on lattice-ordered division rings, duly inspired by an idea of Redfield – McCleary and Schwartz provide answers to questions raised at the conference in problem sessions. McCleary dazzled everybody – again – by solving a problem raised by Holland. Overnight yet. I should also like to quote from Schwartz' paper, because he makes a point in his introduction, which cuts right to the marrow of this thorny political problem of where in mathematics to situate this field of research now deprived of a label:

> ". . . the solution is presented here to advertise a closer cooperation between the groups working in real algebraic geometry and in the theory of lattice-ordered rings."

I am in his debt for deciding to publish his solution in these proceedings.

I am especially glad to have made the acquaintance of Daniel Gluschankof. He practices mathematics in Argentina. He is also typical of hundreds of mathematicians who reside and work in Caribbean or Latin American countries, and who, for reasons which have to do with forces which mathematics is, regretably, powerless to contradict, practice their craft in isolation. The Caribbean Mathematics Foundation, founded shortly before the conference in Curaçao and under whose auspices it took place, seeks to make contacts in the Caribbean and in Latin America with mathematicians like Gluschankof, who have much to offer their counterparts in the so-called developed world. There is every reason to suppose that this acquaintance will be the first in a long series of contacts which will prove beneficial to the mathematical community. I have no misgivings in speaking for Curaçao, the host country for this conference and others to come, when I say that the island is delighted to serve in this capacity. With equal confidence I pronounce the sentence of the participants to the conference held from August the eighth through the twelfth on Curaçao, that the island fulfilled its role admirably.

The composition of the text itself is not, as the reader will soon notice, strictly alphabetical. There are, first, two chapters on the broad themes of groups and rings respectively. Within each chapter the reader will discover a thematic progression; I shall resist characterizing this development, as it is highly subjective, and, in any event, is intended only as an undertone, a kind of basso continuo.

That such a conference as this took place at all is due to the grace and foresight of a number of individuals and the institutions or agencies they represented. I will try to recognize them all, in a reasonably chronological fashion; I apologize for any omissions, as well as for my failure, on occasion, to come up with the adjectives and inflections to thank everybody adequately.

To Lionel Capriles, Managing Director of Maduro's Bank in Curaçao, thanks for pointing me in the right direction at the start of my journey in search of funding for the conference. To Valdemar Marcha, currently the Rector of the University of the Netherlands Antilles, for embracing my proposal with enthusiasm and for sustaining that enthusiasm to this day, enduring gratitude. Likewise, to Charles F. Sidman, Jr., then Dean of the College of Liberal Arts and Sciences of the University of Florida, my gratitude for his support and confidence. To all the 26 speakers at last year's conference a triumphant vote of thanks for their splendid contributions; and to two, in particular, a broader and farther-reaching salute: to Charles Holland, thanks, not only for the advice in preparation for the conference, but for the long years of his friendship as well; and to Paul Conrad for being the catalyst and igniter to all his students, and this one in particular, I take this opportunity to deliver a resounding hurrah on behalf of all of us.

I am always delighted to salute the resourcefulness and spontaneity of my friend Michiel Hazewinkel, whose talents helped make the conference possible.

In my department I was able to count on the support and cooperation of several of my colleagues, notably Al Bednarek, longtime chairman of the department and faithful supporter; Gerard Emch, chairman from 1986 to 1988, who supplied many useful ideas; David Drake, our current chairman, whom I respect enormously for his generosity and his sense of discretion; and, finally, Bruce Edwards, for his friendship and impeccable taste in people and causes.

In Curaçao I was lucky enough to be able to bank on my origins on that island, and to rediscover ancient friendships. My thanks to Agustin Díaz and Ron Gomes-Cásseres for their encouragement; to Grace Jonkhout for her superb judgment and her professionalism; to the Interim Governor of the Netherlands Antilles, Mr. Elmer Joubert, for his valuable advice; to the Ministry of Education of the Antilles and the Department of Education of the island of Curaçao for their warm reception of the concept of the conference, as well as of the participants themselves; to the Princess Beach Hotel, which hosted the conference, for the many ways in which they made our stay one to remember with satisfaction and for a long time to come.

To Sharon Easter of our department, for seeing to it that stipend cheques were delivered on time, thanks; to the staff in our department in general for all the copying, writing and tech-ing, thanks; to the resource people at the University of the Netherlands Antilles who delivered overhead projection equipment and reproduced material to the hotel, thanks.

Finally, to all the authors of this book, my sincerest gratitude for heeding the landmarks I imposed on the calendar, and for being graceful and patient about it. To all referees, thanks for their valuable and time-consuming labor. And to Kluwer Academic Publishers, a salute for deciding to publish this volume, and my thanks for allowing me ample flexibility in shaping it.

Jorge Martínez

Gainesville, June, 1989

ORDERED ALGEBRAIC STRUCTURES
Princess Beach Hotel, Curaçao, August, 1988
Conference Program

Monday, 8 August.

Welcoming ceremonies, Princess Beach Hotel, 8 PM.

Tuesday, 9 August.

Morning Session.

9:00–10:00	**W.A.J. Luxemburg**; California Institute of Technology. *Ordered algebraic structures in analysis.*
10:15–10:45	**C.B. Huijsmans**; Universiteit Leiden. *Almost f-algebras.*
11:00–12:00	**Paul Conrad**; University of Kansas. *A survey of abelian lattice-ordered groups.*

Afternoon Session.

2:30–3:30	**Melvin Henriksen**; Harvey Mudd College. *Rings of continuous functions as a tool in studying lattice-ordered algebraic systems.*
3:45–4:15	**Anthony Hager**; Wesleyan University. *An enrichment of the Yosida representation of an archimedean lattice-ordered group.*
4:30–5:00	**Daniel Gluschankof**; Universidad de Buenos Aires. *Injective objects in the category of abelian lattice-ordered groups with strong order unit.*
5:15–6:15	**Marlow Anderson**; The Colorado College. *Lattice-ordered groups of divisibility.*

Wednesday, 10 August.

Morning Session.

9:00–10:00	**Wayne Powell**; Oklahoma State University. *Universal aspects of lattice-ordered groups.*
10:15–10:45	**Akbar Rhemtulla**; University of Alberta. *Extensions of ordered groups.*
11:00–12:00	**W. Charles Holland**; Bowling Green State University. *Ordered representation theory of groups.*

Afternoon Session.

2:30–3:30	**Steve McCleary**; University of Georgia. *The gate completion of a lattice-ordered group.*
3:45–4:45	**Problem Session I.** (Holland) *An informal discussion of problems in ordered groups, with particular emphasis on the roles of universal algebra and model theory.*
5:00–6:00	**Michael Darnel**; Indiana University at South Bend. *Free lattice-ordered groups over nilpotent groups.*

Evening: Dinner at Indonesia; 8 PM.

Thursday, 11 August.

Morning Session.

8:45–9:45	**Manfred Droste**; Universität Essen. *Automorphism groups of infinite semilinear orders.*
10:00–11:00	**Problem Session II.** (Conrad-Martinez) *An informal discussion of problems in ordered groups.*
11:15–11:45	**John Dauns**; Tulane University. *Partially ordered ∗-division rings.*
12:00–12:30	**Robert Redfield**; Hamilton College. *Embeddings of ordered rings.*

No Afternoon Session. Excursion; approx. 3 hours.

Friday, 12 August.

Morning Session.

8:45–9:45	**Richard Ball**; Boise State University. *Extensions of an archimedean lattice-ordered group defined by the preservation of a designated lattice of ℓ-ideals.*
10:00–11:00	**Problem Session III.** (Henriksen) *An informal discussion of ordered rings, with attention to the role of topology.*
11:15–11:45	**Suzanne Larson**; Loyola Marymount University. *Primary ℓ-ideals in a class of f-rings.*
12:00–12:30	**Constantine Tsinakis**; Vanderbilt University. *The underlying lattice of a lattice-ordered group.*

Friday, 12 August.

Afternoon Session.

2:30–3:00	**Niels Schwartz**; Universität Passau. *Lattice-ordered rings in semi-algebraic geometry.*
3:15–3:45	**James Madden**; Indiana University at South Bend. *Geometry and lattice-ordered groups.*
4:00–5:00	**Problem Session IV.** (Prestel) *An informal discussion of problems in semi-algebraic geometry.*
5:15–5:45	**Alexander Prestel**; Universität Konstanz. *Representations of polynomials as sums of 2-th powers.*
6:00–6:30	**Piotr Wojciechowski**; Warsaw/Bowling Green State University. *Structure of ℓ-nilpotent rings.*

Saturday, 13 August.

Morning Session.

8:45–9:15	**Michèle Giraudet;** Université du Maine. *Cancellation and absorption of lexicographic powers of totally ordered abelian groups.*
9:30–10:00	**F. Lucas;** Université d'Angers. *Some applications of definable spine analysis in ordered abelian groups.*
10:15–11:15	**Problem Session V.** (Luxemburg) *An informal discussion of problems relating to ordered structures occurring in analysis.*
11:30–12:00	**Klaus Keimel;** Technische Hochschule Darmstadt. *Non-archimedean ordered cones and approximation.*
12:15–12:45	**Paulo Ribenboim;** Queen's University. *Commutative convolution rings or ordered semigroups.*

CHAPTER ONE:

GROUPS

LATTICE-ORDERED GROUPS OF DIVISIBILITY: AN EXPOSITORY INTRODUCTION

MARLOW ANDERSON
Department of Mathematics
The Colorado College
Colorado Springs, Colorado 80903
USA

ABSTRACT. *This paper discusses the interplay between the theories of abelian ℓ-groups and Bezout domains via the group of divisibility. This interplay depends on a one-to-one correspondence between onto ℓ-homomorphisms and overrings. A method for interpreting ℓ-embeddings in the context of Bezout domains is conjectured.*

An important application of the theory of abelian partially ordered groups is to the study of the groups of divisibility of integral domains; indeed, this study was of crucial historical importance in the development of the field. The case where the group of divisibility is lattice-ordered is of particular interest and the only one we will consider here. The reader should consult Gilmer [**G**], Môckor [**M**] and Chapter 11 of Anderson and Feil [**AF**] for further mathematical and bibliographic information about this material.

Let D be an integral domain with group of units U, quotient field K and as usual $K^* = K\backslash\{0\}$. Then the ***group of divisibility*** of D (denoted by G or $G(D)$) is the multiplicative group K^*/U; we will abuse notation and write this group additively: $Ua + Ub = Uab$. This group can be equipped with a partial order compatible with the addition provided

$$Ua \leq Ub \Leftrightarrow \frac{b}{a} \in D.$$

Alternatively, consider the set $\{Dx : x \in K^*\}$ of nontrivial cyclic D-submodules of K (also called the ***principal fractional ideals*** of D). We can then define an addition by $Dx + Dy = Dxy$ and a partial order by reverse inclusion. This gives a partially ordered group isomorphic to $G(D)$ via $Ux \mapsto Dx$.

We can now ask for which domains is this partially ordered group an ℓ-group. The answer, as first observed by Jaffard, are the ***pseudo-Bezout*** domains: that is, domains where each pair of nonunits admit a greatest common divisor (which is unique up to multiplication by units); such rings are also called GCD-domains. The proof of Jaffard's result is quite easy: the meet of group elements Ua and Ub is Uc, where c is (any) gcd of a and b.

In the important case where the ℓ-group is actually totally ordered, this association reduces to the classical theory of field valuations. In like spirit, we can view the case of ℓ-groups in this light. Namely, let K be a field; then a ***demivaluation*** on

J. Martinez (ed.), Ordered Algebraic Structures, 3–9.

K is a group homomorphism $w : K^* \to G$ onto an ℓ-group satisfying

$$w(x+y) \geq w(x) \wedge w(y).$$

Then let

$$D_w = \{x \in K : w(x) \geq 0\} \cup \{0\}.$$

It is then easily checked that this is a pseudo-Bezout domain whose group of divisibility is ℓ-isomorphic to G.

It is now natural to ask which abelian ℓ-groups arise as the group of divisibility of some domain. The answer, obtained by Krull in the totally ordered case and independently by Jaffard and Kaplansky in the general case, is that all ℓ-groups do. To prove this, suppose that G is an arbitrary abelian ℓ-group and k is a field. Consider the group ring $k[G]$; this is a domain (since G is torsion free) and so has a quotient field K. Define a map w on $k[G]^*$ into G by setting

$$w\left(\sum a_i X^{g_i}\right) = \bigwedge g_i.$$

This map is then easily extended to K^*; one can then prove that $G(D_w)$ is ℓ-isomorphic to G.

Ohm later observed that the domain constructed above actually satisfies the stronger condition of being a *Bezout* domain: namely, each finitely generated ideal is principal. It is a straightforward exercise to verify that a domain is Bezout if and only if it is pseudo-Bezout and the gcd of a pair of elements can always be written as a linear combination of the elements, as in elementary number theory. Ohm's observation means that the distinction between Bezout and pseudo-Bezout domains is lost upon passage to groups of divisibility, and consequently for our purposes we may as well assume that all domains considered are Bezout.

The crucial piece of machinery necessary to relate the theories of abelian ℓ-groups and Bezout domains is a correspondence theorem we will describe below, between ℓ-homomorphic images of the ℓ-group and overrings of the corresponding Bezout domain. To describe that theorem we need the following terminology from ring theory.

Let D be a domain, K its quotient field, and R any subring of K containing D; we call such R *overrings* of D. For a Bezout domain all overrings are actually *localizations* (see [G]), in the following sense: A set S is called a *multiplicative system* if $0 \notin S$, S is closed under multiplication, and if a divides b and $b \in S$, then $a \in S$ (that is, S is closed under taking divisors). The localization of D at S is then

$$D_S = \left\{\frac{d}{s} \in K : d \in D, s \in S\right\}.$$

The assertion above about Bezout domains means that all overrings occur as D_S, for a (uniquely determined) multiplicative system S. Of course, a case of particular interest is when the multiplicative system $S = D\backslash I$, where I is an ideal; this occurs precisely when the ideal is prime.

We can now state the correspondence theorem between overrings of a Bezout domain and the ℓ-homomorphic images of its group of divisibility:

THEOREM. *For a Bezout domain D there are one-to-one correspondences between these sets:*

(1) *overrings of D;*
(2) *multiplicative systems of D;*
(3) *convex ℓ-subgroups of $G(D)$;*
(4) *ℓ-homomorphic images of $G(D)$.*

Furthermore, $G(D_S) \approx G(D)/H$, where $S = \{d \in D : Ud \in H\}$.

The correspondence between (1) and (2) is the obvious order-reversing one described above, while the correspondence between (3) and (4) is the obvious one between ℓ-homomorphisms and their kernels (which are the convex ℓ-subgroups of $G(D)$). The statement at the end of the theorem describes the order-preserving correspondence between (2) and (4), and also suggests to the reader the straightforward proof required.

In case the multiplicative system is the complement of a (necessarily) prime ideal, localization gives a ***valuation domain***, whose group of divisibility is totally ordered. In ℓ-group theory those convex ℓ-subgroups which are kernels of ℓ-homomorphisms onto totally ordered groups are called prime. The theorem above thus gives an order-reversing correspondence between the prime ideals of the Bezout domain and the prime subgroups of the ℓ-group. This observation is the classical context of the above theorem, as enunciated in various places by Krull, Yakabe, Sheldon and Jaffard.

This correspondence between primes of the domain and primes of the ℓ-group has been extensively explored by many authors, especially by constructing ℓ-groups whose primes satisfy a given property, which then leads to a Bezout domain whose primes satisfy that (order-reversed) property; note that it is usually much easier to construct an ℓ-group whose primes satisfy a given property than to directly construct a Bezout domain. We will mention here only a couple of examples of this approach, to illustrate the principle involved:

(1): The primes of a Bezout domain form a *tree* (that is, a partially ordered set Γ for which $\{\alpha : \alpha \leq \gamma\}$ is totally ordered, for any $\gamma \in \Gamma$). This corresponds to the fact that the primes of an ℓ-group form a *root system* (the dual of a tree). Furthermore, all trees occur as the set of primes of some Bezout domain. This is easy to prove using ℓ-groups, because for any root system Γ, the ℓ-group

$$\Sigma(\Gamma, \mathbf{R}) = \{f : \Gamma \to \mathbf{R} : \text{the support of } f \text{ is finite}\}$$

has a root system of primes isomorphic to Γ.

(2): Every Bezout domain is an intersection of overrings which are valuation domains. This classical result of Krull translates directly to the classical result of Fuchs and Lorenzen that all (abelian) ℓ-groups are subdirect products of totally ordered groups, since the intersection of the overrings of the domain leads to an intersection of convex ℓ-subgroups equalling $\{0\}$.

We shall now take more general advantage of the correspondence theorem. For that purpose, we define a *kernel class* of (abelian) ℓ-groups to be a class $\mathcal{K}$ of abelian ℓ-groups closed under

(K1) ℓ-isomorphisms;
(K2) subdirect products;
(K3) the trivial group belongs to $\mathcal{K}$.

Evidently, every ℓ-group G admits a unique minimum convex ℓ-subgroup $\mathcal{K}(G)$ such that $G/\mathcal{K}(G) \in \mathcal{K}$; that is,

$$\mathcal{K}(G) = \cap\{K : G/K \in \mathcal{K}\},$$

the *$\mathcal{K}$-kernel* of G. (A more general but related concept is that of a *torsion-free class* of ℓ-groups, due to Martinez [Ma].)

We now make a "dual" definition: a *hull class* of Bezout domains is a class $\mathcal{H}$ of Bezout domains closed under

(H1) If $D \in \mathcal{H}$ and $G(D) \approx G(R)$, then $R \in \mathcal{H}$;
(H2) intersection of overrings;
(H3) all fields belong to $\mathcal{H}$.

Evidently, every Bezout domain D admits a least overring $\mathcal{H}(D) \in \mathcal{H}$; that is,

$$\mathcal{H}(D) = \cap\{R : D \subseteq R \subseteq K, R \in \mathcal{H}\},$$

the *$\mathcal{H}$-hull* of D.

We now have the following:

THEOREM.

(1) *Let $\mathcal{H}$ be a hull class, and let*

$$G(\mathcal{H}) = \{\ell\text{-groups } G : G \approx G(D), D \in \mathcal{H}\}.$$

Then $G(\mathcal{H})$ is a kernel class.

(2) *Let $\mathcal{K}$ be a kernel class, and let*

$$D(\mathcal{K}) = \{\text{Bezout domains } D : G(D) \in \mathcal{K}\}.$$

Then $D(\mathcal{K})$ is a hull class.

(3) *The maps G and D are inverse maps which establish a one-to-one correspondence between the collections of kernel and hull classes.*

(4) *Let D be a Bezout domain, $\mathcal{H}$ a hull class, and $\mathcal{K}$ a kernel class. Then*

(i) $\mathcal{K}(G(D)) \approx G(D(\mathcal{K})(D))$;

(ii) $G(\mathcal{H}(D)) \approx G(D)/G(\mathcal{H})(G(D))$; and

(iii) $G(D(\mathcal{K})(D)) \approx G(D)/\mathcal{K}(G(D))$.

An illustrative and important example of this theorem is the correspondence between the class $\mathcal{A}$ of archimedean ℓ-groups and the class $\mathcal{C}$ of completely integrally closed domains. (Recall that an ℓ-group is *archimedean* if $0 \leq ng < h$ for all positive integers n implies that $g = 0$. A domain is *completely integrally closed* if every element of K almost integral over D belongs to D; more explicitly, this means that $a \in D, x \in K^*$ and $ax^n \in D$ for all positive integers n, imply that $x \in D$.) It is a straightforward matter to check that D is completely integrally closed if and only if $G(D)$ is archimedean.

Note that $\mathcal{C}(D)$ is the unique smallest completely integrally closed overring of D; however this need not be the classically defined *complete integral closure* D^c of D. This is precisely because the complete integral closure need not be completely integrally closed. In fact, Heinzer has proved (see [G]) that we can describe D^c in terms of the group of divisibility.

To do this, we require a definition from ℓ-group theory. An element g of an ℓ-group G is *infinitely small* if there exists h for which $n|g| \leq h$, for all positive integers n; the set $B(G)$ of all infinitely small elements of G is a convex ℓ-subgroup. Clearly an ℓ-group is archimedean if and only if $B(G) = \{0\}$.

Given a Bezout domain with complete integral closure D^c and complete integral hull $\mathcal{C}(D)$, Heinzer's result implies the corresponding relationships

$$D \quad \subseteq \quad D^c \quad \subseteq \quad \mathcal{C}(D)$$

and

$$G(D) \rightarrow G(D)/B(G(D)) \rightarrow G(D)/\mathcal{A}(G(D)).$$

The archimedean kernel $\mathcal{A}(G)$ of an ℓ-group G clearly contains all infinitely small elements, but it is not hard to construct an example of an ℓ-group G where the kernel must be larger than $B(G)$ (and with that an example of a Bezout domain whose complete integral closure is not completely integrally closed). The classical construction of the archimedean kernel of an ℓ-group involves considering the concept of "relative uniform closure" (see [**LZ**]); however, it may also be obtained using transfinite induction, by repeatedly considering the convex ℓ-subgroups $B(G)$ of infinitely small elements:

$$B_1(G) = B(G), \quad B_2(G)/B_1(G) = B(G/B(G)), \cdots.$$

For another more general example of the hull-kernel correspondence, let $\mathcal{F}$ be any class of ℓ-groups closed under ℓ-isomorphisms, and let $\mathcal{K}$ be the class of those ℓ-groups which are subdirect products of elements in $\mathcal{F}$. This provides a large class of examples of kernel classes and their corresponding hull classes. For example, if $\mathcal{F} = \{\mathbf{Z}\}$, then $\mathcal{K}$ is the class of ℓ-groups consisting of subdirect products of integers, and the corresponding hull class consists of those domains which are intersections of discrete valuation domains.

We have now seen that all onto ℓ-homomorphisms of an abelian ℓ-group can be intelligently interpreted in terms of groups of divisibility. To completely mirror the theory of ℓ-groups in the theory of Bezout domains, we would thus like to interpret ℓ-embeddings of $G(D)$ into some larger ℓ-group, in terms of the domain D. That is, given a Bezout domain D and ℓ-embedding $G(D) \to H$, can we embed $D \to R$ so that $G(R)$ is ℓ-isomorphic to H in some nice way?

The setup we'd like to use to accomplish this is as follows, where R and D are Bezout domains with corresponding groups of units and quotient fields, and all maps are inclusions:

$$\begin{array}{ccccc} U_R & \longrightarrow & R & \longrightarrow & K_R \\ \uparrow & & \uparrow & & \uparrow \\ U_D & \longrightarrow & D & \longrightarrow & K_D, \end{array}$$

and $K_D \cap R = D$.

It is then easy to show that $K_D^* \cap U_R = U_D$ and so

$$G(D) = K_D^*/U_D = K_D^*/(K_D^* \cap U_R) \approx (K_D^* \times U_R)/U_R \subseteq K_R^*/U_R = G(R).$$

After checking that this map preserves order, we have that $G(D)$ may be considered an ℓ-subgroup of $G(R)$.

An interesting and important example of this situation is the embedding of $\mathbf{Z}$ into the ring I of algebraic integers (which is in fact a Bezout domain; see [**K**]). The group of divisibility $G(\mathbf{Z})$ is a small sum of copies of $\mathbf{Z}$ (one copy for each prime number); the construction above "sparsely" embeds this ℓ-group into $G(I)$, a divisible archimedean ℓ-group with no basis. This embedding in essence encodes classical multiplicative number theory into the language of abelian ℓ-groups.

Unfortunately, it is presently unknown whether every ℓ-group embedding can be so represented. We state this as a conjecture:

CONJECTURE. *Given any Bezout domain D, and any ℓ-embedding of $G(D)$ into an abelian ℓ-group H, it is possible to construct a Bezout domain R as in the diagram above so that H is ℓ-isomorphic to $G(R)$.*

Many nice consequences would flow from this result; various ℓ-group completion theorems (see [**AF**], Chapter 8) would lead to corresponding completion theorems for Bezout domains. For example, every (abelian) ℓ-group can be ℓ-embedded (in a minimal way we shall not make precise here) as an ℓ-subgroup of a *laterally complete* ℓ-group (that is, an ℓ-group where every pairwise disjoint set (of any cardinality) has a least upper bound). Given the conjecture, we could then prove a theorem stating that every Bezout domain could be embedded in a minimal way into a Bezout domain in which every set of elements which is pairwise relatively prime has a least common multiple.

Note that this approach to obtaining theorems about Bezout domains is analogous to Krull's approach to embedding any integrally closed domain into a Bezout domain using the Kronecker function ring, which corresponds to embedding a semiclosed partially ordered group into an ℓ-group (for definitions and discussion see [**G**]).

REFERENCES

[**AF**] Marlow Anderson and Todd Feil, "Lattice-ordered Groups: An Introduction," D. Reidel, Dordrecht, Holland, 1988.

[**G**] Robert Gilmer, "Multiplicative Ideal Theory," Marcel Dekker, New York, 1972.

[**K**] Irving Kaplansky, "Commutative Rings," University of Chicago, Chicago, 1974.

[**LZ**] W.A.J. Luxemburg and A.C. Zaanen, "Riesz Spaces," North Holland, Amsterdam, 1971.

[**Ma**] Jorge Martinez, *The fundamental theorem on torsion classes of lattice-ordered groups*, Trans. AMS **259** (1980), 311–317.

[**M**] Jiri Močkor, "Groups of Divisibility," D. Reidel, Dordrecht, Holland, 1983.

UNIVERSAL ASPECTS OF THE THEORY OF LATTICE ORDERED GROUPS

WAYNE B. POWELL[1]
Oklahoma State University
Stillwater, Oklahoma 74078

To Laszlo Fuchs with deep admiration on the occasion of his 65th birthday.

0. Introduction.

Universal algebra deals with the investigation of classes of algebras. There are essentially two goals of this study. First of all, properties are sought which appear in many classes of algebras. These common threads between different algebras are then investigated in a general setting. Secondly, these properties are studied within specific classes to determine the qualities that are special to the particular classes. These qualities serve as a means of comparing the algebraic structures of the different classes.

In this chapter we consider the applications of universal algebra to ℓ-groups (lattice ordered groups). This presentation will run along the lines of the second goal mentioned above, and so it will primarily involve consideration of universal structures in specific classes of ℓ-groups.

In general an algebra is a set S with various operations, or functions, $* : S^n \longrightarrow S$ $(n \geq 0)$. When viewed as an algebra, an ℓ-group can be thought of as a set G with operations $\cdot, ^{-1}, 1, \wedge$, and $\vee$ such that $\cdot, \wedge$, and $\vee$ are binary, $^{-1}$ is unary, and 1 is nullary. These operations satisfy the laws which require that:

(i) $(G, \cdot, ^{-1}, 1)$ is a group with identity 1.

(ii) $(G, \wedge, \vee)$ is a lattice.

(iii) $x \cdot (y \wedge z) \cdot u = (x \cdot y \cdot u) \wedge (x \cdot z \cdot u)$ and $x \cdot (y \vee z) \cdot u = (x \cdot y \cdot u) \vee (x \cdot z \cdot u)$.

[1]The author gratefully acknowledges the support of a National Science Foundation (EPSCoR) grant during the preparation of this manuscript.

J. Martinez (ed.), Ordered Algebraic Structures, 11–49.

Note that in this definition each of the conditions placed on the set G and its operations can be stated in terms of an equation involving the operations and the elements of G. This property is important in the development of a universal theory for classes of ℓ-groups.

Within a class of algebras, there are several fundamental constructions that are used to determine the strength of the class. First of all, given any algebra A we look for the *subalgebras* of A. These are subsets of A which also form similar algebras under the restricted operations of A. For ℓ-groups the subalgebras are commonly called *ℓ-subgroups*. Thus, an ℓ-subgroup H of an ℓ-group G is a subset of G which is both a sublattice and a subgroup.

Another construction in a class of algebras is formed by considering a collection $\{A_i | i \in I\}$ from the class. We define the *direct product* $\Pi_{i \in I} A_i$ to be the usual Cartesian product of the A_i with all operations being performed componentwise. For a collection of ℓ-groups, the direct product is usually referred to as the *cardinal product*. The order inherited from the lattice operations is determined componentwise by the orders of the cardinal factors (i.e., the A_i). When only two ℓ-groups, G_1 and G_2, are involved in a cardinal product we refer to the *cardinal sum* and write $G_1 \boxplus G_2$. The symbol $\boxplus$ is intended to distinguish this cardinal sum from the frequently occurring group direct sum $G_1 \oplus G_2$ where no order is involved.

Finally, given two algebras A and B with similar operations, a map $\phi : A \longrightarrow B$ is said to be a *homomorphism* if it preserves all the operations of A. If ϕ is a homomorphism of ℓ-groups, then it is named an *ℓ-homomorphism*. This helps distinguish it from the group homomorphisms which commonly play an integral role in the theory of ℓ-groups. Subgroups H of an ℓ-group G which are *convex* become important when considering ℓ-homomorphisms. These subgroups are the ones with the property that $1 < g < h$ for $g \in G$ and $h \in H$ implies that $g \in G$. The kernels of the ℓ-homomorphisms are just the ℓ-subgroups which are normal subgroups and are convex. These correspond to the congruence relations in universal algebra and are called *ℓ-ideals*.

The primary universal notions that help describe classes of ℓ-groups are free algebras, free products of algebras, and amalgamations of algebras. These will be singled out one at a time in sections 2, 3, and 4. They are integrally related concepts, and knowledge about their properties for specific classes of ℓ-groups yields real understanding about what makes ℓ-groups special. The proper context for considering these structures and their properties has proven to be classes of ℓ-groups which form varieties. These classes will be discussed initially in section 1.

Our intent in this presentation is not to prove theorems. They are all proved elsewhere. Rather, we hope to outline the direction that universal algebra investigations have taken in the theory of ℓ-groups. We will review the sequences in which results appeared and discuss the main theorems. References to original papers should provide the interested reader with adequate proofs. Most of the topics in universal algebra related to ℓ-groups have only been considered in relatively recent history. Many problems have been solved in this short time, but with each solution at least two new problems have arisen. We will point out where some of the most glaring open problems exist.

There are several good reference books on universal algebra which can be used to supplement this material. We would recommend either Grätzer [1979], Pierce [1968], or Burris and Sankappanavar [1981].

With regard to ℓ-groups, the most complete reference on varieties is in Reilly [1989], but the proofs of many theorems will have to be sought from their origins. There is no place where the general information on free ℓ-groups is collected. However, when considering only those ℓ-groups which are free in the class of all ℓ-groups, McCleary [1989] gives up to date versions of the more significant theorems. Many of the proofs for theorems on free products and amalgamations are given in Powell and Tsinakis [1989d], [1989e]. However, in these references the proofs are not of all the historical results but rather only for those on the frontiers of some current problems.

1. Varieties of ℓ-groups.

Universal algebra notions are best considered in the context of various classes of algebras. This allows us to look for properties not as much with regard to specific algebras as with respect to the whole class. Further, there are many classes that have similar properties and guarantee nice behavior of the algebras within them. In order to relate these properties we look at the building blocks of a class of algebras.

In studying classes of algebras we want to know when the classes are closed with respect to the formation of (direct) products, subalgebras, or homomorphic images. Interestingly enough, this is related to other conditions on the class.

Each algebra satisfies a set of equations or identities involving the operations used to define it. The collection of all algebras satisfying a given set of equations is called an *equational class*. ℓ-groups in themselves form an equational class and there are many interesting subclasses that can be defined by equations.

Finally, there are often algebras within a class which can be written as subdirect products of other similar algebras only in a way where the projections are isomorphisms. An algebra such as this is called *subdirectly irreducible*. It is characterized by the fact that it has a minimal nontrivial congruence. In the language of ℓ-group theory, we can translate to say that G is subdirectly irreducible iff it has an a minimal, nontrivial ℓ-ideal. The subdirectly irreducibles in a class are the most basic structures in the class and can be used to generate the class.

The relationship between the aforementioned properties of classes is given by the following fundamental theorem.

THEOREM 1.1 (Birkhoff [1935]) Let $\mathcal{U}$ be a class of algebras. Then the following are equivalent:

1. There is a set of equations E such that $\mathcal{U}$ consists precisely of those algebras satisfying the equations from E. ($\mathcal{U}$ is an equational class.)
2. $\mathcal{U}$ is closed with respect to the formation of products, subalgebras, and homomorphic images.
3. $\mathcal{U}$ consists precisely of those algebras which are subdirect products of its subdirectly irreducible members.

A class that satisfies one of the three equivalent conditions from Theorem 1.1 is called a variety. Thus, a class of ℓ-groups forms a variety iff it is defined by a set of equations involving the group and lattice operations iff it is closed with respect to cardinal products, ℓ-subgroups, and ℓ-homomorphic images iff each ℓ-group within is a subdirect product of subdirectly irreducible ℓ-groups from the class. Varieties are important not only because of the three ways of describing them but also because of the kinds of universal structures that they contain.

One of the primary reasons for studying varieties is this existence of various universal structures within them. In particular free algebras and free products of algebras are always known to exist within a variety. Also, many embedding properties make sense when considered with respect to the algebras in a variety. A bonus for varieties of ℓ-groups is that every ℓ-group has a largest convex ℓ-subgroup that is contained in any given variety (Holland [1979]). For now we will deal with the relationship between various varieties of ℓ-groups with the intent of using these correspondences to locate the universal conditions that hold for ℓ-groups.

The use of varieties in the theory of ℓ-groups dates back to the consideration of free ℓ-groups and certain free extensions in the abelian and representable cases by Weinberg

[1963],[1965] and Bernau [1969], [1970], and subsequently by Conrad [1970], [1971] for the general case. The primary incentive for these works was to describe free ℓ-groups in various varieties rather than to consider varieties in and of themselves.

Martinez initiated the study of varieties of ℓ-groups for their own sake in [1974] and to some extent in earlier papers [1972], [1973]. His work involved an investigation of the containment relationships between various varieties and properties of closure under formation of universal objects. This study was followed by a number of papers discussing relationships between varieties and properties of specific varieties. The most influential of these is Glass, Holland, and McCleary [1980] where the collection of all ℓ-group varieties is examined in the spirit of H. Neumann's [1967] book on group varieties. One of the main contributions of this paper is to give a good introduction to the multiplication operation on the set of varieties. Several other papers, in particular Scrimger [1975] and Smith [1980] dealt with the construction of varieties with special properties.

The collection **L** of all ℓ-group varieties forms a lattice in a nice way. The inclusion relation induces a partial order where $\wedge$ and $\vee$ are defined as

$$\mathcal{U} \wedge \mathcal{V} = \mathcal{U} \cap \mathcal{V} \text{ and } \mathcal{U} \vee \mathcal{V} = \cap\{\mathcal{W} | \mathcal{U}, \mathcal{V} \subseteq \mathcal{W}\}.$$

Whereas meets (greatest lower bounds) of varieties are easy to describe, the key to the calculation of joins (least upper bounds) of varieties is found in the following general result known as "Jónsson's Lemma".

THEOREM 1.2 (Jónsson [1967]) If $\mathcal{U}$ and $\mathcal{V}$ are varieties, then the subdirectly irreducible algebras in $\mathcal{U} \vee \mathcal{V}$ are the subdirectly irreducibles that are in $\mathcal{U} \cup \mathcal{V}$.

Jónsson's Lemma is used regularly when determining properties of joins of varieties. Its importance lies in the fact that there are no more subdirectly irreducibles in the smallest variety containing both $\mathcal{U}$ and $\mathcal{V}$ than those which are in either $\mathcal{U}$ or $\mathcal{V}$.

Another general property of the lattice **L** is easily verified.

THEOREM 1.3 **L** is a distributive (in fact dually Brouwerian) lattice.

In order to describe **L** we look for relationships among the different varieties that compose this lattice. There are a number of specific varieties that are of significance in describing these correspondences. Some that will appear regularly in our discussion are listed here in terms of their equational definitions.

$\mathcal{L}$ = all ℓ-groups		
$\mathcal{N}$ = normal-valued ℓ-groups	$\models$	$[(x \vee 1)(y \vee 1)] \wedge [(y \vee 1)^2(x \vee 1)^2] = (x \vee 1)(y \vee 1)$.
$\mathcal{R}$ = representable ℓ-groups	$\models$	$(x \wedge y)^2 = x^2 \wedge y^2$.
$\mathcal{A}$ = abelian ℓ-groups	$\models$	$xy = yx$.
$\mathcal{E}$ = trivial ℓ-groups	$\models$	$x = y$.

The normal-valued ℓ-groups are more easily described as those satisfying the inequality

$$xy \leq y^2x^2$$

for all $x, y > 1$. The representable ℓ-groups are those that can be embedded in a product of totally ordered ℓ-groups. This variety is the largest one where all subdirectly irreducible ℓ-groups are totally ordered.

An opening question in the study of varieties should be: "How many are there?" There can of course be no more than $2^{\aleph_0}$ since there are only finitely many operations in use. This is in fact the precise cardinality of **L**.

THEOREM 1.4 (Kopytov and Medvedev [1977]) There are $2^{\aleph_0}$ varieties of ℓ-groups.

A variety of ℓ-groups that can be defined by a finite number of equations can be defined by a single equation by use of the lattice operations. The significance of the last result lies with the fact that there must be varieties of ℓ-groups that cannot be defined by a finite number of equations.

Reilly [1981] has extended this result to find uncountably many varieties in a specific interval of significance. His construction is based on a Galois connection between varieties of ℓ-groups and fully invariant subgroups of the free group on a countably infinite set.

THEOREM 1.5 (Reilly) There are sets **A** and **B** of varieties between $\mathcal{R}$ and $\mathcal{N}$ each of cardinality $2^{\aleph_0}$ such that **A** is a chain with intersection $\mathcal{R}$ and the elements of **B** are pairwise incomparable.

The most fundamental relationships in the lattice **L** are given below.

THEOREM 1.6 (i) (Weinberg [1965]) $\mathcal{A}$ is the smallest nontrivial variety of ℓ-groups.
(ii) (Holland [1976]) $\mathcal{N}$ is the largest proper variety of ℓ-groups.

The knowledge that $\mathcal{A}$ is the smallest nontrivial variety of ℓ-groups arose from Weinberg's analysis of free abelian ℓ-groups. The key here is the proof that every such ℓ-group is the subdirect product of copies of the integers. The integers are clearly contained in every nontrivial variety and so must every free abelian ℓ-group. An application of this theorem is that every ℓ-group satisfying the equation $xy = yx$ must also satisfy every equation satisfied by any other nontrivial ℓ-group.

Holland's proof that $\mathcal{N}$ is the largest proper variety of ℓ-groups essentially shows that an ℓ-group is normal-valued if it satisfies an equation that is not satisfied by every ℓ-group.

In general, varieties are classified as to those that are representable and those that are not. Another distinction which is helpful is to consider whether a variety is "solvable", which means that it contains only solvable groups. The varieties that cover $\mathcal{A}$ can be identified according to these classifications as we describe below.

Let $\mathbf{Z}wr\mathbf{Z}$ denote the small wreath product of two copies of $\mathbf{Z}$. This groups admits many total orders. Two of particular interest to our discussion here are:

T_1 : $((a_i), n) \geq (0,0)$ iff either $n > 0$, or $n = 0$ and $a_j > 0$ where j is the largest integer i with $a_i \neq 0$.

T_2 : $((a_i), u) \geq (0,0)$ iff either $n > 0$, or $n = 0$ and $a_j > 0$ where j is the smallest integer i with $a_i \neq 0$.

Now, let $M_1 = (\mathbf{Z}wr\mathbf{Z}, T_1)$ and $M_2 = (\mathbf{Z}wr\mathbf{Z}. T_2)$. These totally ordered groups generate solvable varieties $\mathcal{M}_1$ and $\mathcal{M}_2$, respectively, which are contained in $\mathcal{R}$. Another solvable variety in $\mathcal{R}$ of significance is $\mathcal{N}_0$ which is generated by the rank 2 free nilpotent class 2 group with a natural order.

Solvable ℓ-groups outside of $\mathcal{R}$ can be created by using a simple ℓ-automorphism on the product of copies of the integers. In particular for $n \in \mathbf{Z}^+$ define α on $\boxplus_{\mathbf{Z}}\mathbf{Z}$ to be the automorphism induced by $\alpha(a_i) = a_{i+1(\text{mod } n)}$ where a_j is the integer a in the jth component. This automorphism defines a cyclic extension of $\oplus_{\mathbf{Z}}\mathbf{Z}$ which we label S_n . S_n is lexicographically ordered with the order of $< \alpha >$ taking dominance over that of $\oplus_{\mathbf{Z}}\mathbf{Z}$. For each $n \in \mathbf{Z}^+$, the ℓ-group S_n generates a variety $\mathcal{S}_n$.

Theorem 1.7 The varieties $\mathcal{N}_0$, $\mathcal{M}_1$, $\mathcal{M}_2$, and $\mathcal{S}_p$, where p is a prime, are distinct covers of the variety $\mathcal{A}$. In fact these are the only covers of $\mathcal{A}$ containing only solvable groups, and every cover of $\mathcal{A}$ not contained in $\mathcal{R}$ must be solvable.

The fact that $\mathcal{N}_0$, $\mathcal{M}_1$, and $\mathcal{M}_2$ are the only solvable covers of $\mathcal{A}$ inside $\mathcal{R}$ was established by Medvedev [1977]. The varieties $\mathcal{S}_p$ were shown to be covers of $\mathcal{A}$ by Scrimger [1975] and were later shown to be the only ones outside of R in independent works of Darnel [1987], Reilly [1986], and Gurchenkov and Kopytov [1987].

There are other covers of $\mathcal{A}$ (necessarily in $\mathcal{R}$), but no one knows how many. These can be defined by putting appropriate total orders on a free group. In particular Bergman [1984] showed that the free group admits a total order, dubbed a "special order", with the property that

$$1 < x << y ==> x << x^y.$$

His methods are easily adapted to place total orders (called "dual special orders") on free groups with the condition that

$$1 < x << y ==> x^y << x.$$

In so doing he was able to show through an argument of Feil [1985] that there are in fact two covers of $\mathcal{A}$ in $\mathcal{R}$ which are generated by nonsolvable groups. This led to the following theorem which describes these covers in terms of their generators.

Theorem 1.8 (Powell and Tsinakis [1989c]) Let B_1 be the free group on two generators endowed with a special order, and let B_2 be this group with a dual special order. Then B_1 and B_2 generate distinct covers $\mathcal{B}_1$ and $\mathcal{B}_2$ of $\mathcal{A}$.

The most innovative part of the discovery of the covers $\mathcal{B}_1$ and $\mathcal{B}_2$ lies in Bergman's definition of the special orders on general free groups and his subsequent proof that they do indeed define totally ordered groups.

Let us consider only varieties within $\mathcal{R}$ for a while. The number of these, as with all ℓ-group varieties, is at most $2^{\aleph_0}$. In fact this is precisely how many there are.

Theorem 1.9 (Feil [1982]) There are $2^{\aleph_0}$ varieties of representable ℓ-groups.

Feil's result was established by explicitly constructing an uncountable chain of varieties in $\mathcal{R}$. His method was to define sufficiently many cyclic extensions of $\mathbf{R}$ as to generate distinct varieties. In fact the varieties he created are all metabelian, but his techniques can be generalized to get varieties of ℓ-groups which are solvable of any given length.

Other varieties inside of $\mathcal{R}$ which are of fundamental importance are related to the nilpotent groups. In particular for any positive integer n, let

$$\mathcal{N}_n = \ell\text{-groups which are nilpotent groups of class} \leq n.$$

Then each $\mathcal{N}_n$ clearly forms a variety that contains the variety $\mathcal{N}_0$. Further, $\mathcal{N}_n < \mathcal{N}_m$ iff $n < m$.

THEOREM 1.10 (Hollister [1978], Kopytov [1975]) The varieties $\mathcal{N}_n$, $n \geq 1$, are all contained in $\mathcal{R}$.

One further variety which is important in many universal questions about ℓ-groups was first presented by Martinez [1972]. This variety $\mathcal{W}$, called the *weakly abelian* variety, is defined as follows:

$$\mathcal{W} \models (x \vee 1)^2 \wedge (x \vee 1)^y = (x \vee 1)^y.$$

More simply, an ℓ-group is weakly abelian iff $x^2 \geq x^y$ for all $x > 1$. The nice thing about weakly abelian ℓ-groups is that every convex ℓ-subgroup of one is normal. They also form an upper bound for the nilpotent ℓ-groups.

THEOREM 1.11 (Reilly [1983]) Every nilpotent ℓ-group is weakly abelian.

It is not known what varieties, if any, are between $\mathcal{W}$ and $\vee\mathcal{N}_n$.

Turning now to nonrepresentable varieties, we first consider a simple generalization of the abelian equation which proves to be very useful in classifying ℓ-groups. For each $n \in \mathbf{Z}^+$, let

$$\mathcal{L}_n \models x^n y^n = y^n x^n.$$

Although these varieties are easily defined, they can be used to move all the way up the lattice structure of $\mathbf{L}$.

THEOREM 1.12 (Smith [1980]) $\mathcal{N} = \vee\mathcal{L}_n$.

In contrast to the last theorem, Smith [1980] has also shown that $\mathcal{N}$ is (finitely) join irreducible.

A very striking result gives a complete description of the varieties between each $\mathcal{S}_p$ and $\mathcal{L}_p$ where p is prime. The result is originally due to Gurchenkov [1984], but the description here comes from Holland, Mekler, and Reilly [1986].

In particular for $n \in \mathbf{Z}^+$ let $\mathbf{Z}_p = \{0, 1, 2, \ldots, p-1\}$ and denote by Γ_n the set of all k-tuples $(x_1, \ldots, x_k)$ where $x_i \in \mathbf{Z}_p$ and $k \leq n$. Γ_n is partially ordered in such a way as to form a tree (a.k.a. a root system) in the following way:

$$(x_1, \ldots, x_j) \geq (y_i, \ldots, y_i) \text{ iff } x_\ell > y_\ell \text{ for the smallest } \ell \text{ with } x_\ell \neq y_\ell$$

or

$$x_\ell = y_\ell \text{ for } k = 1, \ldots, j \text{ and } j \leq i.$$

Now let G_n be the abelian group of all functions $f : \Gamma_n \longrightarrow \mathbf{Z}$ lattice ordered by $f \geq 0$ if $f(x) \geq 0$ for each maximal element in the support of f. Every automorphism α of Γ_n induces an automorphism of G_n in a natural way. For each $i = 1, 2, \ldots, n$ let α_i be the automorphism of G_n induced by the map

$$(x_i, \ldots, x_k) \longmapsto (x_1, x_2, \ldots, x_i + 1(\text{mod} p), \ldots, x_k).$$

The group of automorphisms generated by $\{\alpha_1, \alpha_2, \ldots, \alpha_n\}$ creates a splitting extension G_n' of G_n. Letting $\mathcal{G}_n$ be the variety generated by G_n' gives the theorem.

THEOREM 1.13 (Gurchenkov [1984]) For any prime p the varieties between $\mathcal{S}_p$ and $\mathcal{L}_p$ are precisely those in the chain

$$\mathcal{S}_p \leq \mathcal{G}_{p^2} \leq \mathcal{G}_{p^3} \leq \cdots \leq \mathcal{G}_{p^n} \leq \cdots \leq \mathcal{L}_p.$$

The determination of the meet irreducible varieties in $\mathbf{L}$ is completed by the next theorem.

THEOREM 1.14 (Powell and Tsinakis [1985b]) The only meet irreducible varieties of ℓ-groups are $\mathcal{E}$, $\mathcal{N}$, and $\mathcal{L}$.

In addition to the lattice operations acting on **L**, there is also a "multiplication" defined by

$G \in \mathcal{U} \cdot \mathcal{V}$ iff there exists $H \in \mathcal{U}$ such that H is an ℓ-ideal of G and $G/H \in \mathcal{V}$.

In H. Neumann's [1967] classic book there is a beautiful development of the effects of the product of varieties of groups. Martinez [1974] first considered the analogue for ℓ-group varieties, and the results of Neumann were mimicked to some degree of completeness by Glass, Holland, and McCleary [1980].

The product and the lattice operations on **L** interact in a nice way.

THEOREM 1.15 (Glass, Holland, and McCleary [1980]) The following rules hold in **L**.

1. $\mathcal{U}(\mathcal{V} \vee \mathcal{W}) = \mathcal{U}\mathcal{V} \vee \mathcal{U}\mathcal{W}$.
2. $(\vee_i \mathcal{U}_i)\mathcal{V} = \vee_i(\mathcal{U}_i\mathcal{V})$.
3. $\mathcal{V}(\wedge_i \mathcal{U}_i) = \wedge_i(\mathcal{V}\mathcal{U}_i)$.
4. $(\wedge_i \mathcal{U}_i)\mathcal{V} = \wedge_i(\mathcal{U}_i\mathcal{V})$.

The product in **L** produces very few idempotents. It also has a freeness determined by the indecomposable varieties. These varieties are the ones that cannot be written in the form $\mathcal{U}\mathcal{V}$ for nontrivial varieties $\mathcal{U}$ and $\mathcal{V}$.

THEOREM 1.16 (a) (Martinez [1974]) The varieties $\mathcal{E}$, $\mathcal{N}$, and $\mathcal{L}$ are idempotent.
(b) (Glass, Holland, and McCleary [1980]) The varieties of ℓ-groups properly contained in $\mathcal{N}$ form a free semigroup generated by the set of indecomposable varieties.

The results of Glass, Holland, and McCleary are proved making extensive use of permutation groups. They further show that the variety $\mathcal{A}$ along with the product defined on **L** can be used to classify varieties as to their relative size as per the next theorem.

THEOREM 1.17 (Glass, Holland, and McCleary [1980]) $\mathcal{N} = \vee_n \mathcal{A}^n$.

2. Free ℓ-groups.

One of the most important structures in many classes of algebras is known as the free algebra. Intuitively such an algebra should be thought of as the loosest way of generating an algebra in the class from a given set using all the operations of the other algebras.

Specifically, let $\mathcal{U}$ be a class of ℓ-groups and let X be a nonempty set. The algebra $F_{\mathcal{U}}(X)$ is called the $\mathcal{U}$-*free* ℓ-group if $F_{\mathcal{U}}(X) \in \mathcal{U}, X$ generates $F_{\mathcal{U}}(X)$ as an ℓ-group, and whenever $H \in \mathcal{U}$ and $\lambda : X \longrightarrow H$ is a map, then there exists an ℓ-homomorphism $\sigma : F_{\mathcal{U}}(X) \longrightarrow H$ which extends λ. Of course it is not always possible to find a $\mathcal{U}$-free ℓ-group in a class of ℓ-groups. There are, however, very important classes that guarantee the existence of these structures.

THEOREM 2.1 (Birkhoff [1935]). If $\mathcal{U}$ is a class of ℓ-groups closed with respect to the formation of cardinal products and ℓ-subgroups, then the $\mathcal{U}$-free ℓ-group on any set $X \neq \emptyset$ will exist in $\mathcal{U}$. Furthermore, this algebra is unique up to isomorphism.

COROLLARY 2.2 $\mathcal{U}$-free ℓ-groups in a variety $\mathcal{U}$ of ℓ-groups exist for any nonempty set X.

The standard way of showing the existence of the $\mathcal{U}$-free ℓ-group in a variety is to construct all possible words using the letters from the set X and the operations $\cdot, ^{-1}, 1, \wedge$, and $\vee$. An equivalence relation is then defined on this set of words by relating those words that can be made equal through the use of some of the equations defining $\mathcal{U}$. The set of equivalence classes forms an element of $\mathcal{U}$ which is in fact the $\mathcal{U}$-free ℓ-group on X.

It is easy to see that every ℓ-group in a variety is the homomorphic image of a free ℓ-group in the variety. This property in itself gives inspiration for an investigation of the structure of these free ℓ-groups. Unfortunately, the construction in terms of words does not shed much light on the specific properties of the free ℓ-groups. Hence, we will look for other methods of describing these algebras in terms of ℓ-group conditions. First of all we consider another related structure.

Let P be a partial ℓ-group. That is, P is a set with partial operations corresponding to the usual ℓ-group operations $\bullet, ^{-1}, 1, \wedge$, and $\vee$ such that whenever the operations are defined for elements of P then the ℓ-group laws are satisfied. We can sometimes freely generate an ℓ-group from P in different classes of ℓ-groups. In particular if $\mathcal{U}$ is a class of ℓ-groups, the *$\mathcal{U}$-free extension* of P in $\mathcal{U}$ is the ℓ-group $F_{\mathcal{U}}(P)$ with the properties that $F_{\mathcal{U}}(P) \in \mathcal{U}$, P generates $F_{\mathcal{U}}(P)$ as an ℓ-group, and whenever $H \in \mathcal{U}$ and $\lambda : P \longrightarrow H$ is a partial homomorphism, there exists an ℓ-homomorphism $\sigma : F_{\mathcal{U}}(P) \longrightarrow H$ extending λ.

As in the case of $\mathcal{U}$-free ℓ-groups, $\mathcal{U}$-free extensions are most often considered within the context of varieties. The reason for this is the next theorem.

THEOREM 2.3 (Grätzer and Schmidt [1963]) Let $\mathcal{U}$ be a class of ℓ-groups closed under ℓ-homomorphic images and the formation of cardinal products. If P is a partial ℓ-group, then $F_{\mathcal{U}}(P)$ exists iff P can be embedded in an ℓ-group in $\mathcal{U}$ using a partial homomorphism.

Free extensions can be constructed using words in a fashion similar to the method used for free algebras. The only difference is in the equivalence relation defined on the set of words. In the present case we also identify words that can be made equal by use of the partial operations of P. Thus, if P is a set with no operations, the free extension will agree with the free algebra. If P is already an ℓ-group, then its free extension is itself.

Free extensions have unfortunately received little attention in the theory of ℓ-groups outside of one special case. Suppose P is a partially ordered group. Then P has implicit partial operations $\wedge$ and $\vee$ as determined by the partial order. Using these together with the full group operations, P can be considered as a partial ℓ-group. When we refer to the free extension of a partially ordered group it is the free extension of this partial ℓ-group to which we will actually be addressing ourselves.

In several early papers on $\mathcal{U}$-free ℓ-groups (especially Weinberg [1963], [1965], Bernau [1969], [1970], Conrad [1970], and Powell [1981]) the concept of a free ℓ-group over a partially ordered group was exploited. This notion relates to our definition of a $\mathcal{U}$-free extension with one important note. When a free ℓ-group over a partially ordered group P (which might possibly be an ℓ-group already) was calculated in these early papers, then the partial operations considered were not all of the ones implicit from the order relation. Rather, the only operations on P taken into consideration were those between elements related to each other. That is to say, $x \wedge y$ and $x \vee y$ exist in the partial ℓ-group P iff $x \leq y$ or $y \leq x$. To distinguish this special case from the general free extension we will use the symbol P' to represent the partial algebra created from a partially ordered group by using only the lattice operations that arise from comparable elements.

The distinction made in the preceding paragraph is important. If for example P is an ℓ-group in $\mathcal{U}$, then $F_{\mathcal{U}}(P) = P$. However, $F_{\mathcal{U}}(P')$ will be P only when P is totally ordered.

$\mathcal{U}$-free extensions of partial ℓ-groups have been most frequently considered for the classes $\mathcal{A}$ and $\mathcal{L}$, although many of the techniques and results that have been achieved can be extended to other varieties, in particular to $\mathcal{R}$ and the nilpotent varieties. We will concentrate here on the two main varieties where the research was originally directed keeping in mind that extensions of results are possible.

There are several descriptions of free ℓ-groups which yield important structural theorems. Initially, Weinberg [1963], [1965] considered these structures for $\mathcal{A}$. His approach was to give a general description of the $\mathcal{A}$-free extension $F_{\mathcal{A}}(P')$ and find a special case that corresponded to the $\mathcal{A}$-free ℓ-group. Along with his construction several important structural results followed. Some of the original proofs were lacking and the results were improved and corrected by Bernau [1969], [1970]. The general approach used by Weinberg and Bernau is appealing because it easily generalizes to other varieties inside $\mathcal{R}$, and in fact to $\mathcal{L}$. It also is of significance because of its application to more general free extensions.

Two other descriptions of $\mathcal{A}$-free ℓ-groups have been useful. In Conrad [1971] a description was given of the generators of the $\mathcal{A}$-free ℓ-group inside a cardinal product of copies of the integers. Further, Baker [1968] gave a very elegant description of these free ℓ-groups in topological terms. His approach does not appear to be easily extendable beyond the abelian variety, but it does have nice realizations for vector lattices. It also yields simple proofs of some difficult algebraic results. We will concentrate first on the techniques of Weinberg in $\mathcal{A}$ and their generalizations to $\mathcal{L}$. We will also give a brief description of the topological approach of Baker.

Initially, review of Theorem 2.3 begs the question of when free extensions exist for varieties of ℓ-groups. Here the notion of a *semiclosed* partially ordered group comes into play. P is such a group if $x \in P^+$ whenever $x \in P$ and $x^n \in P^+$ for some $n \in \mathbf{Z}^+$.

THEOREM 2.4 (a) (Weinberg [1963]) Let P be a partially ordered, torsion-free abelian group. Then $F_{\mathcal{A}}(P')$ exists iff P is semiclosed.
(b) (Conrad [1970]). Let P be a partially ordered, torsion-free group. Then $F_{\mathcal{L}}(P')$ exists iff P is the intersection of the positive cones of right orders on P.

To achieve a description of the $\mathcal{U}$-free extensions in A let P be a partially ordered, torsion-free abelian group. Then set

$$\Gamma = \{T | T \quad \text{is the positive cone of a total order on } P \text{ containing } \ P^+\}.$$

With a natural partial ℓ-homomorphism we can embed P' in the product $\Pi_\Gamma(P,T)$ where the product is taken over all total orders in Γ. P' becomes a partial subalgebra of this product. As each of the components of the product is a totally ordered group the image of P' generates a sublattice of the product in a natural way. This image is the $\mathcal{A}$-free extension of P'.

THEOREM 2.5 Let P be a semiclosed, partially ordered, torsion-free abelian group. Then the sublattice of $\Pi_\Gamma(P,T)$ generated by the image of P under the usual diagonal map is $F_{\mathcal{A}}(P')$.

The method of this past theorem can be extended to $\mathcal{L}$ by using right orders instead of total orders. In this direction let P be a torsion-free, partially ordered group where the positive cone P^+ is the intersection of right orders on P. If R_α is P with one such right order, then denote by $A(R_\alpha)$ the ℓ-group of order preserving permutations of R_α. Each $x \in P$ corresponds to an element ρ_x of R_α defined by

$$\rho : \ z \longmapsto zx.$$

Let

$$\Lambda = \{A(R_\alpha) | R_\alpha = P \text{ with a right order containing } P^+ \ \}.$$

Then P can be embedded in $\Pi_\Lambda A(R_\alpha)$ by the diagonal map.

THEOREM 2.6 (Conrad [1970]) Let P be a torsion-free, partially ordered group such that P^+ is the intersection of right orders. The $\mathcal{L}$-free extension of P is the sublattice generated by the image of P in $\Pi_\Lambda A(R_\alpha)$ under the diagonal map.

Theorems 2.5 and 2.6 give descriptions of the $\mathcal{U}$-free ℓ-groups in $\mathcal{A}$ and $\mathcal{L}$ if the proper partial ℓ-groups are chosen. For any variety $\mathcal{U}$ of ℓ-groups, let $G(\mathcal{U})$ denote the class of all groups that can be lattice ordered in such a way that they are admitted to $\mathcal{U}$. The class $G(\mathcal{U})$ is closed under the formation of subgroups and group direct products (in fact, it is a quasi-variety of groups) so it contains $G(\mathcal{U})$-free groups. We will consider these groups as partially ordered groups with the trivial order. The following theorem was proved by Weinberg [1963] and Conrad [1970] for the special cases of $\mathcal{U} = \mathcal{A}$ and $\mathcal{U} = \mathcal{L}$, respectively.

THEOREM 2.7 Let $\mathcal{U}$ be a variety of ℓ-groups, and let X be any nonempty set. The $\mathcal{U}$-free ℓ-group on X is the $\mathcal{U}$-free extension of the $G(\mathcal{U})$-free group on X with trivial order.

Thus, the $\mathcal{A}$-free ℓ-group on a set X is constructed in the following way. First form $G = \oplus_X \mathbf{Z}$, the free abelian group on the set X. Endow G with the trivial order. Consider the set $\Gamma = \{T | T \text{ is a total order on } G\}$. Then $F_{\mathcal{A}}(X)$ is the sublattice of $\Pi_\Gamma(G, T)$ generated by the image of G under the diagonal map.

The following diagram represents the universal properties of the free algebras involved in the construction of the $\mathcal{A}$-free ℓ-groups.

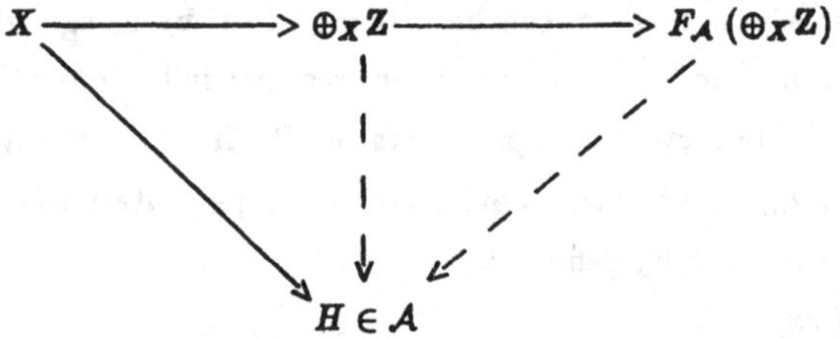

For each of the classes $\mathcal{A}$ and $\mathcal{L}$ the representations described above can be improved considerably. However, the general method mentioned here is sufficient to prove a number of results on the structure of these free algebras. A couple of nice improvements on these representations are given below.

THEOREM 2.8 (Weinberg [1963]) The $\mathcal{A}$-free ℓ-group on a set X can be represented as the sublattice generated by the free abelian group on X inside of a cardinal product of copies of $\mathbf{Z}$.

The representation of Conrad was greatly improved by Kopytov.

THEOREM 2.9 (Kopytov [1979]) The $\mathcal{L}$-free ℓ-group on a set X can be embedded in a natural way into $A(G_X)$ where G_X is the free group on X with a suitable right order.

An alternate approach to $\mathcal{A}$-free ℓ-groups was developed by Baker [1968]. In his work the main emphasis was in analyzing the free vector lattices, but his techniques also carry over to $\mathcal{A}$. The primary theorem of Baker is the following.

THEOREM 2.10 (Baker) Let X be a nonempty set. Then $F_{\mathcal{A}}(X)$ is isomorphic to the ℓ-group of continuous, piecewise linear functions from $\mathbf{R}^{|X|}$ to $\mathbf{R}$ which is generated by the coordinate projections under pointwise operations.

This theorem gives a very nice picture of the free generators of the $\mathcal{A}$-free ℓ-groups. It also gives simple proofs to some of the algebraic results mentioned below. Another advantage is its applicability to vector lattices and ordered modules (see Madden [1983]). Its drawback is that there does not appear to be an analogue for nonabelian varieties of ℓ-groups except possibly in the nilpotent case.

The ability to represent $\mathcal{A}$-free and $\mathcal{L}$-free ℓ-groups yields a solution to the word problem. The abelian case follows from Weinberg's [1963] work while the general case is due to Holland and McCleary [1979].

THEOREM 2.11 The word problems for $F_{\mathcal{A}}(X)$ and $F_{\mathcal{L}}(X)$ are solvable.

Whether or not the general conjugacy problem in $F_{\mathcal{L}}(X)$ is solvable is unkown. However, Arora and McCleary [1986] have shown the solvability for special cases.

Further descriptions of $\mathcal{L}$-free ℓ-groups have been given using the extensive theory of ℓ-permutations groups. A nice, complete treatment of these groups is found in Glass [1981]. Some elementary definitions are as follows. A group $G \subseteq A(X)$ of order preserving permutations of the totally set X is called $o-2-$*transitive* if for all $\alpha < \beta$ and $\lambda < \delta$ in X, there is some $g \in G$ with $g(\alpha) = \lambda$ and $g(\beta) = \delta$. G is called *pathological* if for each nonidentity permutation $g \in G$ the set $\{x \in X | g(x) \neq x\}$ is unbounded in X.

The next theorem describes the $\mathcal{L}$-free ℓ-group in terms of fundamental properties of ordered permutation groups. The finite case was handled by McCleary [1985a] and most of the infinite case by Glass [1974].

THEOREM 2.12 (Glass, McCleary) The $\mathcal{L}$-free ℓ-group on a nonempty set X can be embedded as a pathological, o-2-transitive subgroup of $A(Y)$ for some totally ordered set Y. If X is countable, then Y can be taken to be the set $\mathbf{Q}$ of rational numbers with the usual order.

The last two theorems have been superseded by a more recent result which solidifies the place of $\mathcal{L}$-free ℓ-groups in terms of groups of order preserving permutations.

THEOREM 2.13 (McCleary [1985b]) (GCH) For any nonempty set X such that $|X|$ is regular, there exists a right ordering of the free group G on X such that $F_{\mathcal{L}}(X)$ can be embedded as a pathological, o-2-transitive subgroup of $A(G)$.

There are a number of results giving descriptions of the structure of $\mathcal{U}$-free ℓ-groups. Most, but not all of these, arise from the use of one of the representation theorems mentioned above. We will list the most significant ones here.

A search for projectives within a nontrivial variety $\mathcal{U}$ of ℓ-groups naturally begins by describing these objects as the retracts of the $\mathcal{U}$-free ℓ-groups. It would of course be nice if these retracts were all (cardinal) summands of free ℓ-groups. This is too much to expect, though, beyond the simplest cases. In particular, it is not difficult to see that the $\mathcal{U}$-free ℓ-group on the set $X = \{x\}$ is just $\mathbf{Z} \boxplus \mathbf{Z}$ with x corresponding to the generator $(1, -1)$. However, no other $\mathcal{U}$-free ℓ-groups can be so decomposed.

THEOREM 2.14 (Weinberg, Bernau, McCleary, and Medvedev) The $\mathcal{U}$-free ℓ-group on a set X with $|X| > 1$ cannot be decomposed into a nontrivial cardinal sum.

The preceding theorem was first stated and "proved" by Weinberg [1965] for $\mathcal{A}$-free ℓ-groups using his representation theory. Conrad [1966] subsequently pointed out that the proof was in error. Bernau [1969] then presented another proof (the original version of which was also incorrect, but which was corrected by Weinberg before publication) to finally give the result for $\mathcal{A}$. The final result as presented by Bernau has the additional appeal that it holds for $\mathcal{A}$-free extensions $F_{\mathcal{A}}(P')$ of partially ordered groups P. The case for $\mathcal{L}$-free ℓ-groups was established by McCleary [1985a] using the pathological, o-2-transitive representation of $F_{\mathcal{L}}(X)$. The result for the other varieties was filled-in by Medvedev [1981] by finding suitable epimorphisms of $\mathcal{U}$-free ℓ-groups to $\mathcal{A}$-free ℓ-groups.

Descriptions of the kinds of elements and their frequency of occurrence have been quite successful as the next series of theorems testifies.

THEOREM 2.15 (Weinberg; Powell and Tsinakis; Baldwin, Berman, Glass, and Hodges) Every set of pairwise disjoint elements in the $\mathcal{U}$-free ℓ-group on a set X is countable. Furthermore, the $\mathcal{U}$-free ℓ-group must have a countably infinite set of pairwise disjoint elements if $|X| > 1$.

Again the original result limiting the size of sets of pairwise disjoint elements is due to Weinberg [1965] for the variety $\mathcal{A}$. Subsequently, Powell and Tsinakis [1983a] pointed out an extension of his techniques to arbitrary varieties of ℓ-groups. Simultaneously, and independently, Baldwin et al. [1982] applied a combinatorial argument and the "Δ-system lemma" to get the same result. Finally, Powell and Tsinakis [1984] showed that there do in fact exist infinite sets of this kind in any $\mathcal{U}$-free ℓ-group on a set of more than one element.

At the same time that Powell and Tsinakis [1983a] and Baldwin et al. [1982] were considering sets of disjoint elements, they were also determining the size of maximal chains in the $\mathcal{U}$-free ℓ-groups. Powell and Tsinakis extended the techniques of Galvin and Jónsson [1961] where a similar investigation was conducted for the free lattices while Baldwin et al. again made application of the Δ-system lemma. The result was the following theorem.

THEOREM 2.16 (Powell and Tsinakis; Baldwin, Berman, Glass, and Hodges) Every infinite chain in a $\mathcal{U}$-free ℓ-group must be countable.

We conclude with a series of interesting results on the behavior of the elements of $F_{\mathcal{L}}(X)$.

THEOREM 2.17 (a) (Medvedev [1981]) The center of $F_{\mathcal{L}}(X)$ is trivial.

(b) (McCleary [1985a]) $F_{\mathcal{L}}(X)$ is not a completely distributive lattice.

(c) (Kopytov [1983]) If $x^n = y^n$ in $F_{\mathcal{L}}(X)$, then $x = y$.

(d) (McCleary [1985a]) $F_{\mathcal{L}}(X)$ has no basic elements.

The techniques used to analyze $\mathcal{U}$-free ℓ-groups can also be applied to classes of free ordered modules. Bigard [1973] considered f-modules (lattice ordered modules that are subdirect products of totally ordered modules) over totally ordered left Ore domains. He was successful in applying the direct analogue of Weinberg's methods to the free modules in this class. More generally Powell [1978], [1981] investigated the class of f-modules over very general lattice ordered rings. In this class a construction is given of the free f-modules in the same style as Weinberg's construction for $\mathcal{A}$-free ℓ-groups except that it is necessary to use total orders on *quotients* of free modules. Several structural results are also established by Powell for free f-modules, including their cardinal indecomposability. Madden [1983] also constructed free f-modules using techniques similar to Baker's and gave an alternate proof of cardinal indecomposability. Disjoint sets in free f-modules were considered by Powell and Tsinakis [1989a]. Surprisingly enough their size is not limited as it is in the case of $\mathcal{U}$-free ℓ-groups. In fact for any cardinal m there is a set X with $|X| = m$ such that the free f-module on X has a disjoint set of cardinality m.

3. Free products of ℓ-groups.

Along the same lines as the free algebras, there are structures in many classes that generate algebras in the class from previously existing ones. When this is done in the loosest way possible, the result is known as the free product of the existing algebras.

In particular let $\mathcal{U}$ be a class of ℓ-groups and let G_1 and G_2 be ℓ-groups in $\mathcal{U}$. The *$\mathcal{U}$-free product* $G_1 {}^{\mathcal{U}}\bigsqcup G_2$ of G_1 and G_2 is the ℓ-group in $\mathcal{U}$ containing G_1 and G_2 with the properties:

(i) $G_1 {}^{\mathcal{U}}\bigsqcup G_2$ is generated as an ℓ-group by $G_1 \cup G_2$;

(ii) whenever $\phi_1 : G_1 \longrightarrow H$ and $\phi_2 : G_2 \longrightarrow H$ are ℓ-homomorphisms for $H \in \mathcal{U}$, there exists an ℓ-homomorphism $\psi : G_1 {}^{\mathcal{U}}\bigsqcup G_2 \longrightarrow H$ such that ψ extends both ϕ_1 and ϕ_2.

The ℓ-groups G_1 and G_2 are called the *free factors* of $G_1 {}^{\mathcal{U}}\bigsqcup G_2$. The definition is easily generalized to the free product of any collection of ℓ-groups.

As with $\mathcal{U}$-free ℓ-groups it is not always possible to construct a $\mathcal{U}$-free product of arbitrary ℓ-groups in any class. For classes with sufficient structure, however, the existence of these free products is guaranteed.

THEOREM 3.1 (Christensen and Pierce [1959]) Let $\mathcal{U}$ be a class of ℓ-groups which is closed with respect to the formation of cardinal products and ℓ-subgroups. Then the $\mathcal{U}$-free product of any number of ℓ-groups in $\mathcal{U}$ exists in $\mathcal{U}$.

It follows immediately that $\mathcal{U}$-free products exist in varieties of ℓ-groups. Because of the additional structure of these classes it is natural to restrict consideration of $\mathcal{U}$-free products to these special classes.

$\mathcal{U}$-free products can be constructed by the use of words in a fashion similar to the construction of $\mathcal{U}$-free ℓ-groups. This is done by forming all possible words using the alphabet $G_1 \cup G_2$ and identifying those that are equivalent under the equations defining $\mathcal{U}$ or that can be related by collapsing adjacent elements of the same free factor. The equivalence classes under this identification form a member of $\mathcal{U}$ in a natural way, and it can be seen that this ℓ-group is in fact the $\mathcal{U}$-free product $G_1 {}^{\mathcal{U}}\bigsqcup G_2$.

Intuitively the $\mathcal{U}$-free product of $G_1, G_2 \in \mathcal{U}$ is the element of $\mathcal{U}$ that combines all of G_1 and G_2 with as little interaction among the elements of the free factors as is possible to still allow $G_1 {}^{\mathcal{U}}\bigsqcup G_2 \in \mathcal{U}$. It is readily seen that every ℓ-group in $\mathcal{U}$ which is generated by G_1 and G_2 must be an ℓ-homomorphic image of $G_1 {}^{\mathcal{U}}\bigsqcup G_2$.

Although the investigation of free products of such structures as lattices and groups has a long history, they were not studied within the context of ℓ-groups until Martinez introduced them in [1972] for the class of all ℓ-groups and subsequently for the variety of abelian ℓ-groups in [1973]. His main emphasis was to consider when the $\mathcal{U}$-free product of ℓ-groups G_1 and G_2 in $\mathcal{V} \subset \mathcal{U}$ was contained in $\mathcal{V}$. In fact this seldom occurs.

In Holland and Scrimger [1972] there is also a look at $\mathcal{L}$-free products. It is shown here that the group free product of two ℓ-groups need not be a subgroup of their $\mathcal{L}$-free product. Along these same lines Franchello [1978] looked for and found a connection between the $\mathcal{L}$-free product of ℓ-groups and their free product as distributive lattices.

Most of the results on free products of ℓ-groups in varieties contained in $\mathcal{R}$ are based on a representation scheme presented in Powell and Tsinakis [1983]. This paper gives a general procedure for representing $\mathcal{A}$-free products in the spirit of Weinberg's construction for $\mathcal{A}$-free ℓ-groups. Several special representations allow for some solid structure results on $\mathcal{A}$-free products. This method of representation is generalized to $\mathcal{R}$ in Powell and Tsinakis [1984] and can in fact be used for any variety contained in $\mathcal{R}$.

Beyond the variety $\mathcal{R}$ it is necessary to look at $\mathcal{L}$-free products. In this case Glass [1984] and [1987] has shown that they can be represented in a nice way in terms of groups of permutations of totally ordered sets.

We will outline here the construction of the $\mathcal{A}$-free product since it has produced the most results. The generalization to other representable varieties is left to outside reading. (In particular we defer to Powell and Tsinakis [1984].)

Let $G_1, G_2 \in \mathcal{A}$. Then we consider subgroups K of $G_1 \oplus G_2$ which admit total orders extending the orders of the factors G_1 and G_2 In particular let

$\Lambda = \{((G_1 \oplus G_2)/K, T) \mid K$ is a subgroup of $G_1 \oplus G_2$ and T is the positive cone of a total order extending the orders of G_1 and $G_2\}$.

There is a natural embedding ψ of $G_1 \oplus G_2$ in $\Pi_\Lambda((G_1 \oplus G_2)/K, T)$ and the sublattice generated by the image of this embedding is the free product in $\mathcal{A}$

THEOREM 3.2 (Powell and Tsinakis [1983]) Let $G_1, G_2 \in \mathcal{A}$ and let Λ and ψ be as defined above. Then $G_1 {}^{\mathcal{A}}\bigsqcup G_2$ is the sublattice of $\Pi_\Lambda((G_1 \oplus G_2)/K, T)$ generated by $\psi(G_1 \oplus G_2)$. That is,

$G_1 {}^{\mathcal{A}}\bigsqcup G_2 \cong \{\vee_{i \in I} \wedge_{j \in J} \psi(g_{ij}) | g_{ij} \in G_1 \oplus G_2$ and I and J are finite $\}$.

There are several further representations of the $\mathcal{A}$-free product that are sometimes more useful than the initial description. In particular as a special case we consider the $\mathcal{A}$-free product of totally ordered abelian groups.

THEOREM 3.3 (Martinez [1972]) Let $\mathcal{U}$ be a variety of ℓ-groups and let $G_1, G_2 \in \mathcal{U}$ be totally ordered. Then

$$G_1 {}^{\mathcal{U}}\bigsqcup G_2 \cong F_{\mathcal{U}}(P')$$

where $P = G_1 + G_2$. In particular $G_1 {}^{\mathcal{A}}\bigsqcup G_2$ is the naturally generated sublattice of $\Pi_{\Lambda'}(G_1 + G_2, T)$ where

$\Lambda' = \{((G_1 \oplus G_2), T) | T$ is the positive cone of a total order on $G_1 \oplus G_2$ extending the orders of both G_1 and G_2 $\}$.

The ability to eliminate the consideration of quotients of $G_1 \oplus G_2$ for totally ordered abelian group G_1 and G_2 in the last theorem makes life much easier when dealing with this free product. Beyond the case where the free factors are totally ordered, it becomes much more difficult to limit the number of components used in the product making up the representation of $G_1 {}^{\mathcal{A}}\bigsqcup G_2$. However, the next theorem has proven a major step in this direction. Here we consider quotients of the free factors rather than of the whole direct product. These can be thought of as special quotients of the direct product in a natural way.

THEOREM 3.4 (Powell and Tsinakis [1983]) Let $G_1, G_2 \in \mathcal{A}$ and let

$$\Lambda_0 = \{((G_1/K_1 \oplus G_2/K_2), T) \mid K_1 \text{ is a prime subgroup of } G_1 \text{ and } T \text{ is the positive cone of a total order on } G_1/K_1 \oplus G_2/K_2 \text{ extending the orders of } G_1/K_1 \text{ and } G_2/K_2\}.$$

If $\psi : G_1 \oplus G_2 \longrightarrow \Pi_{\Lambda_0}((G_1/K_1 \oplus G_2/K_2), T)$ is the natural embedding, then $G_1 \,{}^{\mathcal{A}}\!\bigsqcup G_2$ is isomorphic to the sublattice of $\Pi_{\Lambda_0}((G_1/K_1 \oplus G_2/K_2), T)$ generated by $\psi(G_1 \oplus G_2)$.

A further restriction on the number of factors in this large product can be made by limiting the set of quotients to only those involving minimal primes of the free factors.

Theorem 3.3 established the relationship between $\mathcal{U}$-free products of totally ordered groups and $\mathcal{U}$-free extensions. A more general theorem for the class of abelian ℓ-groups also exists.

THEOREM 3.5 (Powell and Tsinakis [1983]). Let $G_1, G_2 \in \mathcal{A}$ and let

$$\Gamma_i = \{G_i/K_i \mid K_i \text{ is a prime subgroup of } G_i\}, i = 1,\ 2.$$

Further, let

$$\Delta = \{F_{\mathcal{A}}(G_1/K_1 \oplus G_2/K_2) \mid G_i/K_i \in \Gamma_i\},$$

and denote by ψ the natural embedding of $G_1 \oplus G_2$ in $\Pi_\Delta(F_{\mathcal{A}}(G_1/K_1 \oplus G_2/K_2))$. Then the sublattice generated by $\psi(G_1 \oplus G_2)$ is isomorphic to the $\mathcal{A}$-free product $G_1 \,{}^{\mathcal{A}}\!\bigsqcup G_2$.

There is one final description of $\mathcal{U}$-free products that has been very beneficial in determining their structure. This one is for the class $\mathcal{L}$ of all ℓ-groups. A special case of this theorem appears in Glass [1984] with the complete result in Glass [1987].

THEOREM 3.6 (Glass) Let G_1 and G_2 be countable ℓ-groups. Then $G_1 \,{}^{\mathcal{L}}\!\bigsqcup G_2$ can be embedded in $A(\mathbf{Q})$ as an o-2-transitive group of permutations.

In considering free products of ℓ-groups it is important to know what happens to the ℓ-subgroups of the free factors when the free products are formed. In this direction we say that a variety $\mathcal{U}$ of ℓ-groups satisfies *the special amalgamation property* if for any $H_1, H_2, G_1, G_2 \in \mathcal{U}$ with $H_1 \subset G_1$ and $H_2 \subset G_2$, the ℓ-subgroup of $G_1 \,{}^{\mathcal{U}}\!\bigsqcup G_2$ generated by $H_1 \cup H_2$ is isomorphic to $H_1 \,{}^{\mathcal{U}}\!\bigsqcup H_2$. In papers on ℓ-groups, this property has sometimes been referred to as the *subalgebra property* for free products, but the current terminology is more consistent with universal algebra literature.

Using Theorem 3.4 the following theorem can be proved.

THEOREM 3.7 (Powell and Tsinakis [1983]) The variety $\mathcal{A}$ satisfies the special amalgamation property.

No other variety of ℓ-groups is known to satisfy the special amalgamation property. However, a number of varieties can be shown to fail this property.

THEOREM 3.8 (Powell and Tsinakis [1983a], [1985a], [1989e], and Powell [1989]) Varieties which fail the special amalgamation property include the nilpotent varieties $\mathcal{N}_n (n > 1)$, the varieties $\mathcal{L}_n (n > 1)$, and an uncountable collection of varieties between $\mathcal{R}$ and $\mathcal{N}$ having intersection $\mathcal{R}$. Further, any variety which is the nontrivial join of varieties must fail the special amalgamation property.

The failure of the special amalgamation property was established initially for the $\mathcal{L}_n$ by a simple argument involving the defining equations. For $\mathcal{N}_n$ this failure is based on the failure of the amalgamation in conjunction with the satisfaction of the congruence extension property. This correspondence will be discussed in the next section. The uncountable collection of varieties containing $\mathcal{R}$ comes from those mentioned in Theorem 1.5. Their failure of the special amalgamation property is again based on a consideration of the defining equations. Varieties of lattices that are $\vee$-reducible were first shown to fail the special amalgamation property by Joel Berman. The proof for ℓ-group varieties is an adaption of this result.

In spite of the failure of this property in general for most varieties, there are many interesting subclasses of a variety where special amalgamation can occur. We refer the reader to Powell [1989] for a discussion of these.

We return now to the relationship of ℓ-group free products to those of the underlying group and lattice structure. As was mentioned previously, Holland and Scrimger have shown that the group free product of two ℓ-groups need not be a subgroup of their $\mathcal{L}$-free product. However, there is a nice relationship between the lattice free product in a certain class of lattices.

Let $\mathcal{D}$ be the class of all distributive lattices with a distinguished element (denoted by e). Then any ℓ-group can be considered as a member of $\mathcal{D}$ by identifying the group identity with e. The class $\mathcal{D}$ is a variety and so admits free products. The relationship of these free products to the ℓ-group free products is striking.

THEOREM 3.9 (Franchello [1978], Powell and Tsinakis [1983b]) Let $\mathcal{U}$ be either of $\mathcal{A}$ or $\mathcal{L}$. If $G_1, G_2 \in \mathcal{U}$, then the $\mathcal{D}$-free product of G_1 and G_2 is the naturally generated sublattice of $G_1 {}^{\mathcal{U}}\bigsqcup G_2$.

The last theorem was initially proved by Franchello for $\mathcal{L}$ using words in the free products involved. The proof for the abelian case was patterned after Franchello's methods by Powell and Tsinakis and relied on the representation theory of Theorems 3.2 and 3.4.

A famous theorem of Baer and Levi [1936] states that no group can be simultaneously decomposed into a free product and a direct product. However, this is not true for abelian groups since the free product in the class of abelian groups agrees with the direct product. For ℓ-group varieties the result of Baer and Levi carries over.

THEOREM 3.10 (Powell and Tsinakis, Glass) Let $\mathcal{U}$ be a variety of ℓ-groups with $G \in \mathcal{U}$. If $\mathcal{U} = \mathcal{A}$ or $\mathcal{L}$, then G cannot be decomposed into both a nontrivial cardinal product and a nontrivial free product. Further, if G is finitely generated, then such decompositions cannot hold in any variety $\mathcal{U}$.

The preceding theorem was first established for $\mathcal{A}$ in Powell and Tsinakis [1982] using Theorem 3.5. This theorem was later extended in Powell and Tsinakis [1984] to all proper varieties $\mathcal{U}$ under the finitely generated restriction. The case for $\mathcal{L}$ was handled by Glass [1987] using Theorem 3.6.

A couple of comments on this last theorem are in order. First of all the $\mathcal{U}$-free ℓ-group on a set X is just the free product of $|X|$ copies of the $\mathcal{U}$-free ℓ-group on one generator (i.e. $\mathbf{Z} \boxplus \mathbf{Z}$). Hence, $\mathcal{U}$-free ℓ-groups are $\mathcal{U}$-free products of finitely generated ℓ-groups and Theorem 2.14 follows as a corollary. Secondly, whether or not the finitely generated restriction can be removed for general varieties $\mathcal{U}$ remains an open problem, but it seems likely that the result will hold in general.

Theorem 3.6 can be used to establish another result based on the properties of o-2-transitive permutations groups.

THEOREM 3.11 (Glass [1987]) The $\mathcal{L}$-free product of two nontrivial ℓ-groups has trivial center.

A further consideration of the properties of elements in free products is inspired by Theorem 2.15 which says that $\mathcal{U}$-free ℓ-groups cannot have sets of disjoint elements of cardinality greater than $\aleph_0$. For free products the problem becomes more complex.

An ℓ-group G will be said to satisfy the m-disjointness condition for the cardinal m if every set S of pairwise disjoint elements has cardinality less than m. Thus, $\mathcal{U}$-free ℓ-groups satisfy the $\aleph_1$-disjointness condition. For $\mathcal{U}$-free products the question is whether or not the m-disjointness condition is preserved by the free product operation.

Adams and Kelly [1977] have extensively examined the preservation of the m-disjointness conditions by free products in varieties of lattices. Among other results they conclude that these conditions are always preserved by free products in the variety of all lattices. Also, if m is a singular or weakly compact cardinal, then the m-disjointness condition is satisfied by free products in the variety of distributive lattices if it is first satisfied by the free factors. The contrasting result for varieties of ℓ-groups is radically different.

THEOREM 3.12 (Powell and Tsinakis [1986], [1988]) Let $\mathcal{U}$ be any variety of ℓ-groups and let m be any cardinal. Then there exist totally ordered ℓ-groups G_1 and G_2 in $\mathcal{U}$ such that $G_1 {}^{\mathcal{U}}\bigsqcup G_2$ fails the m-disjointness condition.

This theorem was first proved for $\mathcal{A}$ and then extended to all varieties of ℓ-groups. In fact the groups in Theorem 3.12 can be chosen to be abelian. In any case this result demonstrates that $\mathcal{U}$-free products fail to preserve the m-disjointness condition in the worst possible fashion since totally ordered groups possess the 3-disjointness condition.

The consideration of disjointness conditions can be used to point out the significance of the operations involved in free products of algebras. For example the real numbers $\mathbf{R}$ may be interpreted as a distributive lattice, an abelian ℓ-group, or a vector lattice. The distributive lattices with distinguished element form a variety $\mathcal{D}$ with operations $\wedge$, $\vee$, and e. The operations in $\mathcal{A}$ are of course $\wedge, \vee, +, -$, and 0. The vector lattices also form a variety of algebras by adding to the operations of $\mathcal{A}$ those of the scalar multiplication of each element from $\mathbf{R}$. These varieties are related by

$$\mathcal{V} \subset \mathcal{A} \subset \mathcal{D},$$

although the operations of each are different.

It follows from Adams and Kelly [1977] that $\mathbf{R}\,^{\mathcal{D}}\!\bigsqcup \mathbf{R}$ has no infinite set of pairwise disjoint elements. On the other hand Bleier [1975] has shown that the free vector lattice on any number of generators satisfies the $\aleph_1$-disjointness condition. The free vector lattice on one generator is $\mathbf{R}\boxplus\mathbf{R}$ and as a consequence of Theorem 3.7 applied to vector lattices it follows that $\mathbf{R}\,^{\mathcal{V}}\!\bigsqcup \mathbf{R}$ also satisfies the $\aleph_1$-disjointness condition. Further, an extension of Powell and Tsinakis [1984] to vector lattices shows that in fact there is an infinite disjoint set in this free product. Interestingly enough, the $\mathcal{A}$-free product $\mathbf{R}\,^{\mathcal{A}}\!\bigsqcup \mathbf{R}$ has an uncountable set of pairwise disjoint elements (Powell and Tsinakis [1986]).

There are other results on free products in Martinez [1972], [1973] which are of interest. In particular it is shown there that $\mathcal{A}$-free products have no "basic" elements and no singular elements that are strictly positive in one of the free factors. Further consideration is given to characterizing when $\mathcal{A}$-free products can be archimedean. This latter condition is given another look in Bixler et al. [1989] for subfields of $\mathbf{R}$. The final theorem of this section gives some results on placing free products from one class in a smaller class.

THEOREM 3.13 (Martinez [1972]) Let G_1 and G_2 be nontrivial ℓ-groups. Then

(i) $G_1\,^{\mathcal{L}}\!\bigsqcup G_2 \notin \mathcal{R}$;

(ii) if $G_1, G_2 \in \mathcal{N}$, then $G_1\,^{\mathcal{N}}\!\bigsqcup G_2 \notin \mathcal{R}$;

(iii) if $G_1, G_2 \in R$, then $G_1\,^{\mathcal{L}}\!\bigsqcup G_2 \notin \mathcal{N}$.

There are still a number of open questions on free products in various classes of ℓ-groups. In addition to filling in the gaps in some of the hypotheses of the theorems given here, a few problems are virtually untouched. It is unknown whether or not free products satisfy the refinement property. There is little known about the preservation of the size of maximal chains by free products. In this case an investigation along the same lines as that for disjointness conditions is in order. Another topic with a multitude of problems is to determine how to freely decompose an ℓ-group in a particular variety as well as to determine when such a decomposition is unique. Finally, as a consequence of the representation theorems for $\mathcal{A}$-free products and their generalizations to $\mathcal{R}$-free products (Powell and Tsinakis [1984]) there is a real need to learn more about the possibilities of extending partial orders on groups to total orders.

Free products have also been considered in other classes of ordered algebraic structures. In Madden [1983] a topological method of describing the A-free products is achieved as a special case of a construction of free products of lattice ordered modules. A development of a representation theory of lattice ordered modules similar to the one described for $\mathcal{A}$-free products in this section is given in Cherri and Powell [1989b]. Most, but not all, of the results in $\mathcal{A}$ carry over to appropriate classes of ordered modules generated by totally ordered modules. There is quite a bit of room for further work here as the rings involved play an intricate role, and the various conditions placed on them can have a major effect on the structure of the free product (witness the situation for disjointness conditions of free products in $\mathcal{A}$ and $\mathcal{V}$).

4. The amalgamation property.

The notion of amalgamating two algebras arises naturally from group theory where there has been extensive research on combining two groups in such a way as to preserve a common subgroup. This is usually done through a free product with amalgamated subgroup. This concept became of universal interest after Jónsson [1956], [1960] began a general investigation of classes of algebras where some form of amalgamation is guaranteed.

Specifically, a class $\mathcal{U}$ of ℓ-groups is said to have the *amalgamation property* if given any quintuple $\{A, B_1, B_2, \alpha_1, \alpha_2\}$ with $A, B_1, B_2 \in \mathcal{U}$ and with $\alpha_1 : A \longrightarrow B_1$ and $\alpha_2 : A \longrightarrow B_2$ being embeddings, there exists a triple $\{C, \beta_1, \beta_2\}$ with $C \in \mathcal{U}$ and with $\beta_1 : B_1 \longrightarrow C$ and $\beta_2 : B_2 \longrightarrow C$ being embeddings such that $\beta_1\alpha_1 = \beta_2\alpha_2$. The quintuple $\{A, B_1, B_2, \alpha_1, \alpha_2\}$ is called a V-*formation* in $\mathcal{U}$ and the triple $\{C, \beta_1, \beta_2\}$ is called an *amalgam* of this quintuple.

The amalgamation property is often depicted by the following diagram.

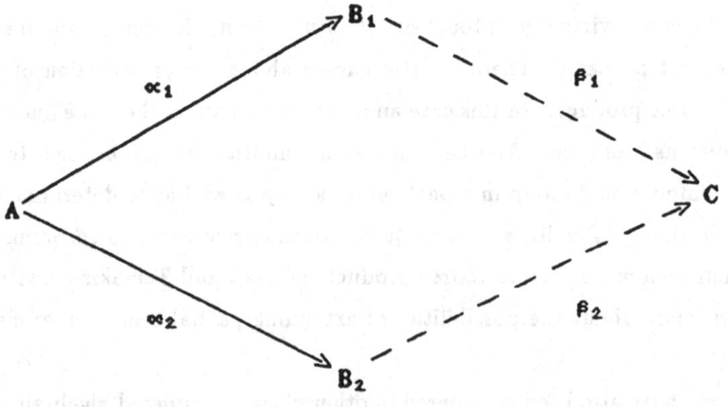

Problems concerning the amalgamation property for classes of ℓ-groups have been primarily restricted to varieties because of their relationship to free products. The next theorem is well-known for groups, but was not stated for ℓ-groups until Cherri and Powell [1989a]. Here we look at a variety $\mathcal{U}$ of ℓ-groups with $B_1, B_2 \in \mathcal{U}$. Given the quintuple $Q = \{A, B_1, B_2, \alpha_1, \alpha_2\}$ in $\mathcal{U}$ we consider the free product $B_1 {}^{\mathcal{U}}\bigsqcup B_2$ and the natural embeddings $\lambda_1 : B_1 \longrightarrow B_1 {}^{\mathcal{U}}\bigsqcup B_2$ and $\lambda_2 : B_2 \longrightarrow B_1 {}^{\mathcal{U}}\bigsqcup B_2$. Denote by N the ℓ-ideal of $B_1 {}^{\mathcal{U}}\bigsqcup B_2$ generated by all words of the form $\lambda_1\alpha_1(a)\lambda_2\alpha_2(a)^{-1}$ where $a \in A$. Then there is a natural projection $\sigma : B_1 {}^{\mathcal{U}}\bigsqcup B_2 \longrightarrow (B_1 {}^{\mathcal{U}}\bigsqcup B_2)/N$.

THEOREM 4.1 If $\mathcal{U}$ is a variety of ℓ-groups, then the quintuple $\{A, B_1, B_2, \alpha_1, \alpha_2\}$ can be amalgamated in $\mathcal{U}$ iff it can be amalgamated by the triple $\{C, \beta_1, \beta_2\}$ where $C = B_1 {}^{\mathcal{U}}\bigsqcup B_2/N$ and β_1 and β_2 are the natural maps from B_1 and B_2 to C.

Thus, free products are integrally related to amalgamations, and the logical setting for investigating this property is again in classes which are varieties.

The investigation of amalgamations of ℓ-groups was begun by Pierce [1972a], [1972b], [1976]. In this early work the primary emphasis is on determining if the amalgamation is satisfied by $\mathcal{L}$ and $\mathcal{A}$. A construction is given of the amalgam (previously shown to exist) in $\mathcal{A}$. There is also an investigation of classes other than varieties that satisfy this property. The next plunge into questions on amalgamations came in Powell and Tsinakis [1985], [1989b] and Glass, Saracino, and Wood [1984] where it is shown that a number of varieties fail the amalgamation property.

Other problems on amalgamations have been in the direction of related topics such as consideration of ℓ-subgroups of free products and relating amalgamation to embedding an ℓ-group in a divisible ℓ-group.

Our first information will be on relating the amalgamation property in a variety to other properties of the variety. Initially, a variety $\mathcal{U}$ is said to have the *congruence extension property* if whenever $G \in \mathcal{U}$ and H is an ℓ-subgroup of G, then every ℓ-ideal I of H can be extended to an ℓ-ideal J of G in such a way that $I = J \cap H$.

The first part of the next theorem is due to Jónsson [1961] while the second part involving the congruence extension property comes from Grätzer and Lakser [1971].

THEOREM 4.2 (Jónsson; Grätzer and Lakser) Let $\mathcal{U}$ be a variety of ℓ-groups. If $\mathcal{U}$ satisfies the amalgamation property, then $\mathcal{U}$ also satisfies the special amalgamation property. If the congruence extension property holds in $\mathcal{U}$, then $\mathcal{U}$ satisfies the amalgamation property iff it satisfies the special amalgamation property.

The amalgamation property is also closely related to the ability to embed an ℓ-group in a variety in a divisible member of that variety. In particular we say that a variety $\mathcal{U}$ of ℓ-groups satisfies the *divisible embedding property* if any member of $\mathcal{U}$ can be embedded in a divisible member of $\mathcal{U}$. The correspondence between this property and amalgamations is given in the next theorem.

THEOREM 4.3 (Powell and Tsinakis [1985a]) Let $\mathcal{U}$ be a variety of ℓ-groups. Then $\mathcal{U}$ satisfies the divisible embedding property iff every V-formation $\{\mathbf{Z}, \mathbf{Q}, B_2, \alpha_1, \alpha_2\}$ in $\mathcal{U}$ can be amalgamated in $\mathcal{U}$.

The divisible embedding property can in turn be related to the special amalgamation property by the following theorem.

THEOREM 4.4 (Powell and Tsinakis [1989e]) Let $\mathcal{U}$ be a variety of ℓ-groups. If $\mathcal{U}$ satisfies the divisible embedding property, then for each $G \in \mathcal{U}$ the ℓ-subgroup of $\mathbf{Q}^{\mathcal{U}} \bigsqcup G$ generated by $\mathbf{Z} \cup G$ is in fact $\mathbf{Z}^{\mathcal{U}} \bigsqcup G$.

The congruence extension property, the special amalgamation property, and the divisible embedding property can all be used to establish the validity or nonvalidity of the amalgamation property in varieties of ℓ-groups.

Varieties that satisfy the congruence extension property include $\mathcal{A}$ and the nilpotent varieties $\mathcal{N}_n$. Thus, the satisfaction of the special amalgamation property in $\mathcal{A}$ yields immediately that $\mathcal{A}$ also satisfies the amalgamation property (see Theorem 3.7). This fact was originally proved by Pierce [1972a] without the aid of free products and Theorem 4.2.

The sequence of implications for the varieties $\mathcal{N}_n$ is reversed from that for $\mathcal{A}$. It was shown by Powell and Tsinakis [1985a] that none of these varieties have the amalgamation property and hence we get a portion of Theorem 3.8 using Theorem 4.2.

The failure of the amalgamation property has been shown for many other varieties of ℓ-groups. Initially Pierce [1972a] showed the failure of this property for the variety $\mathcal{L}$. Implicit in his proof was the failure of amalgamation in varieties containing $\mathcal{A}^2$. Also, Powell and Tsinakis [1985a] considered an uncountable collection of varieties between $\mathcal{R}$ and $\mathcal{N}$. The intersection of these varieties is $\mathcal{R}$, they form a chain, and they contain all of the varieties $\mathcal{S}_p$ (p prime).

Furthermore, it is shown here that any variety contained in one of these varieties and not containing a nonabelian representable ℓ-group must fail the divisible embedding property and hence also the amalgamation property.

Inside $\mathcal{R}$ failure of the amalgamation property is rampant. As was mentioned above the nilpotent, nonabelian varieties fail this property. Further, Glass, Saracino, and Wood [1984] proved that $\mathcal{R}$ itself fails the condition. Their proof of failure also included the weakly abelian variety and could easily be adapted to shown failure in many other unidentified varieties. Powell and Tsinakis [1989b] extended these results to show that any variety in one of the intervals $[\mathcal{M}_1, \mathcal{R}]$ and $[\mathcal{M}_2, \mathcal{R}]$ must fail the amalgamation property. The proof given there for $\mathcal{R}$ was also much simpler than the original one. Further, included in the intervals mentioned is the uncountable collection of varieties first defined by Feil [1982].

The results on the satisfaction or lack thereof of the amalgamation property are summarized in the following theorems.

THEOREM 4.5 (Pierce) $\mathcal{A}$ satisfies the amalgamation property.

THEOREM 4.6 The following varieties of ℓ-groups fail the amalgamation property.

(i) (Pierce) $\mathcal{L}$.

(ii) (Pierce) All varieties $\mathcal{U}$ such that $\mathcal{A}^2 \subseteq \mathcal{U}$.

(iii) (Powell and Tsinakis) An uncountable chain $\mathcal{U}_\alpha$ of varieties having intersection $\mathcal{R}$. Also, all varieties $\mathcal{V}$ satisfying $\mathcal{R} \cap \mathcal{V} = \mathcal{A}$ and $\mathcal{V} \subset \mathcal{U}_\alpha$ (some α). These include each $\mathcal{S}_p$ and each $\mathcal{L}_n$.

(iv) (Powell and Tsinakis) The nilpotent varieties $\mathcal{N}_n (n > 1)$.

(v) (Glass, Saracino, and Wood; Powell and Tsinakis) All varieties in the intervals $[\mathcal{M}_1, \mathcal{R}]$ and $[\mathcal{M}_2, \mathcal{R}]$.

(vi) (Glass, Saracino, and Wood) The variety $\mathcal{W}$ of weakly abelian ℓ-groups.

For varieties of lattices, Day and Jezek [1984] have shown that any proper variety containing the variety of distributive lattices (which happens to be the smallest nontrivial lattice variety) must fail the amalgamation property. It is unlikely that one all-encompassing proof of the analogous result will be found for varieties of ℓ-groups. However, it is also unlikely that any nonabelian variety of ℓ-groups does satisfy the amalgamation property. The main gaps in completing this classification are the varieties outside of $\mathcal{R}$ which are not contained in any of the varieties discussed in Powell and Tsinakis [1985a] and also the varieties which contain nonsolvable ℓ-groups (e.g. $\mathcal{B}_1$ and $\mathcal{B}_2$). There is also a small gap around the nilpotent varieties since the result there does not hold for $\mathcal{N}_0$ or any possible varieties containing all $\mathcal{N}_n$ and properly contained in $\mathcal{W}$.

Questions on amalgamation in varieties of ℓ-groups more appropriately take on the form of determination of which V-formations $\{A, B_1, B_2, \alpha_1, \alpha_2\}$ can be amalgamated. This is more reasonable than trying to show that all V-formations in the class can be amalgamated. Along these lines we consider the *amalgamation class*, Amal($\mathcal{U}$), of a variety $\mathcal{U}$. This consists of all $A \in \mathcal{U}$ with the property that every V-formation $\{A, B_1, B_2, \alpha_1, \alpha_2\}$ in $\mathcal{U}$ can be amalgamated in $\mathcal{U}$. This class can in fact be quite small as the next theorem testifies.

THEOREM 4.7 (i) (Pierce [1972a]) If $\mathcal{U}$ is a variety of ℓ-groups such that $\mathcal{A}^2 \subseteq \mathcal{U}$, then $\mathcal{A}$ is not contained in Amal ($\mathcal{U}$).
(ii) (Powell and Tsinakis [1985a]) No nontrivial, totally ordered abelian group is contained in Amal($\mathcal{N}_n$) ($n > 1$).

A variation of the amalgamation property that has proved fruitful in other classes of algebras is known as the *strong amalgamation property*. For this property to be satisfied by a class $\mathcal{U}$ we require that for every V-formation $\{A, B_1, B_2, \alpha_1, \alpha_2\}$ in $\mathcal{U}$ there is an amalgam $\{C, \beta_1, \beta_2\}$ in $\mathcal{U}$ such that $\beta_1\alpha_1 = \beta_2\alpha_2$ and further that $\beta_1(b_1) = \beta_2(b_2)$ implies that $b_1 \in \alpha_1(A)$ and $b_2 \in \alpha_2(A)$. Jónsson [1960] has shown that the variety of all lattices satisfies this strong amalgamation property. Due to the failure of amalgamation in $\mathcal{L}$ the analogous result cannot hold for ℓ-groups. In fact it cannot even come close.

THEOREM 4.8 (Cherri and Powell [1989a]) Every variety in the interval $[\mathcal{A}, \mathcal{R}]$ fails the strong amalgamation property.

If the embeddings of A in B_1 and B_2 (in $\mathcal{A}$) are performed in such a way as to make the images of A convex, then not only can the V-formation be amalgamated, but it can also be done strongly.

THEOREM 4.9 (Cherri and Powell [1989a]) A V-formation $\{A, B_1, B_2, \alpha_1, \alpha_2\}$ in $\mathcal{A}$ can be strongly amalgamated in $\mathcal{A}$ if $\alpha_1(A)$ and $\alpha_2(A)$ are convex in B_1 and B_2, respectively.

In addition to determination of the validity of the amalgamation property in the remaining varieties, a number of other possible problems remain. It would be of greater use to describe the amalgamation class of varieties rather than just showing that the general property fails. In particular giving a description of this class for $\mathcal{R}$ would be of special significance. Further results along the lines of Theorem 4.9, where specific V-formations are identified as having amalgams in the given variety, would allow for positive applications of this important notion.

References

[1977] M. E. Adams and D. Kelly, "Disjointness conditions in free products of lattices", *Algebra Universalis* 7(1977), 245–258.

[1986] A. K. Arora and S. H. McCleary, "Centralizers in free lattice-ordered groups", *Houston J. Math.* 12(1986), 455–482.

[1936] R. Baer and F. Levi, "Freie products und ihre Untergruppen", *Comp. Math.* 3 (1936), 391–398.

[1968] K. Baker, "Free vector lattices", *Can. J. Math.* 20(1968), 58–66.

[1982] J. T. Baldwin, J. Berman, A. M. W. Glass, and W. Hodges, "A combinatorial fact about free algebras", *Algebra Universalis* 15(1982), 145–152.

[1984] G. Bergman, "Specially ordered groups", *Comm. Alg.* 12(19)(1984), 2315–2333.

[1969] S. Bernau, "Free abelian lattice groups", *Math. Ann.* 180(1969), 48–59.

[1970] S. Bernau, "Free non-abelian lattice groups", *Math. Ann.* 186(1970), 249–262.

[1935] G. Birkhoff, "On the structure of abstract algebras", *Proc. Camb. Phil. Soc.* 31(1935), 433–454.

[1989] P. Bixler, P. Conrad, W. B. Powell, and C. Tsinakis, "Torsion classes of vector lattices", *J. Austral. Math. Soc. Series A*, 1989.

[1975] R. D. Bleier, "Free ℓ-groups and vector lattices", *J. Austral. Math. Soc.* 19(1975), 337–342.

[1989a] M. Cherri and W. B. Powell, "Strong amalgamations of lattice ordered groups and modules", 1989.

[1989b] M. Cherri and W. B. Powell, "Free products of lattice ordered modules", preprint.

[1981] S. Burris and H. P. Sankappanavar, *A Course in Universal Algebra*, Springer-Verlag, New York, 1981.

[1959] D. J. Christensen and R. S. Pierce,"Free products of α-distributive Boolean algebras", *Math. Scand.* 7(1959), 81–105.

[1969] P. Conrad, *Math Reviews*, 31(1966), 1054.

[1970] P. Conrad, "Free lattice-ordered groups", *J. Algebra* 16(1970), 191–203.

[1971] P. Conrad, "Free abelian ℓ-groups and vector lattices", *Math. Ann.* 190(1971), 306–312.

[1987] M. Darnel, "Special-valued ℓ-groups and abelian covers", *Order* 4 (1987), 191–194.

[1984] A. Day and J. Jezek, "The amalgamation property for varieties of lattices", *Trans. Amer. Math. Soc.* 286(1984), 251–256.

[1982] T. Feil, "An uncountable tower of ℓ-group varieties", *Algebra Universalis* 14(1982), 129–131.

[1985] T. Feil, "Varieties of representable ℓ-groups", *Ordered Algebraic Structures*, Marcel Dekker, New York, 1985.

[1978] J. D. Franchello, "Sublattices of free products of lattice ordered groups", *Algebra Universalis* 8(1978), 101–110.

[1974] A. M. W. Glass, "ℓ-simple lattice-ordered groups", *Proc. Edinburgh Math. Soc.* 19(1974), 133–138.

[1981] A. M. W. Glass, *Ordered Permutation Groups*, Cambridge University Press, London, 1981.

[1984] A. M. W. Glass, "Effective embeddings of countable lattice-ordered groups", *Proc. 1st International Sym. on Ordered Algebraic Structures*, Helderman-Verlag, 1984, 63–69.

[1987] A. M. W. Glass, "Free products of lattice-ordered groups", *Proc. Amer. Math. Soc.* 101(1987), 11–16.

[1980] A. M. W. Glass, W. C. Holland, and S. McCleary, "The structure of ℓ-group varieties", *Algebra Universalis* 10(1980), 1–20.

[1984] A. M. W. Glass, D. Saracino, and C. Wood, "Non-amalgamation of ordered groups", *Math. Proc. Camb. Phil. Soc.* 95(1984), 191–195.

[1979] G. Grätzer, *Universal Algebra, 2nd ed.*, Springer-Verlag, New York, 1979.

[1971] G. Grätzer and H. Lakser, "The structure of pseudocomplemented distributive lattices. II: congruence extension and amalgamation", *Trans. Amer. Math. Soc.* 156(1971), 343–358.

[1963] G. Grätzer and E. T. Schmidt, "Characterizations of congruence lattices of abstract algebras", *Acta. Sci. Math.* 24(1963), 34–59.

[1984] S. A. Gurchenkov, "Varieties of ℓ-groups with the identity $[x^P, y^P] = e$ have finite bases", *Algebra and Logic* 23(1984), 20–35.

[1987] S. A. Gurchenkov and V. M. Kopytov, "On the description of the coverings of a variety of abelian lattice-ordered groups", *Sibirsk. Mat. Zh.* 28(1987), 66-69.

[1976] W. C. Holland, "The largest proper variety of lattice-ordered groups", *Proc. Amer. Math. Soc.* 57(1976), 25–28.

[1979] W. C. Holland, "Varieties of ℓ-groups are torsion classes", *Czech. Math. J.* 29(1979), 11–12.

[1979] W. C. Holland and S. McCleary, "Solvability of the word problem in free lattice-ordered groups", *Houston J. Math.* 5(1979), 99–105.

[1986] W. C. Holland, A. H. Mekler, and N. R. Reilly, "Varieties of lattice ordered groups in which prime powers commute", *Algebra Universalis* 23(1986), 196–214.

[1972] W. C. Holland and E. Scrimger, "Free products of lattice-ordered groups", *Algebra Universalis* 2(1972), 247–254.

[1978] H. A. Hollister, "Nilpotent ℓ-groups are representable", *Algebra Universalis* 8(1978), 65–71.

[1956] B. Jónsson, "Universal relational systems", *Math. Scand.* 4(1956), 193–208.

[1960] B. Jónsson, "Homogeneous universal relational systems", *Math. Scand.* 8(1960), 137–142.

[1961] B. Jónsson, "Sublattices of a free lattice", *Can. J. Math.* 13(1961), 256–264.

[1967] B. Jónsson, "Algebras whose congruence lattices are distributive", *Math. Scand.* 21(1967), 110–121.

[1975] V. M. Kopytov, "Lattice-ordered locally nilpotent groups", *Algebra i Logika* 14(1975), 407–413.

[1979] V. M. Kopytov, "Free lattice-ordered groups", *Algebra and Logic* 18(1979), 259–270.

[1983] V. M. Kopytov, "Free lattice-ordered groups", *Siberian Math. J.* 24(1983), 98–101.

[1977] V. M. Kopytov and N. I. Medvedev, "Varieties of lattice-ordered groups", *Algebra and Logic* 16(1977), 281–285.

[1983] J. Madden, "Two methods in the study of K-vector lattices", doctoral dissertation, Wesleyan University, 1983.

[1972] J. Martinez, "Free products in varieties of lattice-ordered groups", *Trans. Amer. Math. Soc.* 172(1972), 249–260.

[1973] J. Martinez, "Free products of abelian ℓ-groups", *Czech. Math. J.* 23(98)(1973), 349–361.

[1974] J. Martinez, "Varieties of lattice-ordered groups", *Math. Zeit.* 137(1974), 265–284.

[1985a] S. H. McCleary, "Free lattice-ordered groups represented as o-2-transitive ℓ-permutation groups", *Trans. Amer. Math. Soc.* 290(1985), 69–79.

[1985b] S. H. McCleary, "An even better representation for free lattice-ordered groups", *Trans. Amer. Math. Soc.* 290(1985), 81–100.

1989] S. H. McCleary, "Free lattice-ordered groups", *Lattice-Ordered Groups*, Reidel, 1989.

1977] N. Y. Medvedev, "The lattices of varieties of lattice-ordered groups and Lie algebras", *Algebra and Logic* 16(1977), 27–30.

1981] N. Y. Medvedev, "Decomposition of free ℓ-groups into ℓ-direct products", *Siberian Math. J.* 21(1981), 691–696.

1967] H. Neumann, *Varieties of Groups*, Springer-Verlag, New York, 1967.

1972a] K. R. Pierce, "Amalgamations of lattice ordered groups", *Trans. Amer. Math. Soc.* 172(1972) 249–260.

1972b] K. R. Pierce, "Amalgamating abelian ordered groups", *Pac. J. Math.* 42(1972), 711–723.

1976] K. R. Pierce, "Amalgamated sums of abelian ℓ-groups", *Pac. J. Math.* 65(1976), 167–173.

1968] R. S. Pierce, *Introduction to the Theory of Abstract Algebras*, Holt, Rinehart, and Winston, New York, 1968.

1978] W. B. Powell, "On a class of lattice ordered modules", doctoral dissertation, Tulane University, 1978.

1981] W. B. Powell, "Projectives in a class of lattice ordered modules", *Algebra Universalis* 13(1981), 24–40.

1989] W. B. Powell, "Special amalgamation of lattice ordered groups", preprint.

1982] W. B. Powell and C. Tsinakis, "Free products of abelian ℓ-groups are cardinally indecomposable", *Proc. Amer. Math. Soc.* 86(1982), 385–390.

1983a] W. B. Powell and C. Tsinakis, "Free products in the class of abelian ℓ-groups", *Pacific J. Math.* 104(1983), 429–442.

1983b] W. B. Powell and C. Tsinakis, "The distributive lattice free product as a sublattice of the abelian ℓ-group free product", *J. Austral. Math. Soc.* 34(1983), 92–100.

1984] W. B. Powell and C. Tsinakis, "Free products of lattice ordered groups", *Algebra Universalis* 18(1984), 178–198.

[1985a] W. B. Powell and C. Tsinakis, "Amalgamating lattice ordered groups", *Ordered Algebraic Structures,* Marcel Dekker, New York, 1985.

[1985b] W. B. Powell and C. Tsinakis, "Meet irreducible varieties of lattice ordered groups", *Algebra Universalis* 20(1985), 262–263.

[1986] W. B. Powell and C. Tsinakis, "Disjointness conditions for free products of ℓ-groups", *Arkiv. der Math.* 46(1986), 491–498.

[1988] W. B. Powell and C. Tsinakis, "Disjoint sets in free products of representable ℓ-groups", *Proc. Amer. Math. Soc.* 104(1988), 1014–1020.

[1989a] W. B. Powell and C. Tsinakis, "Disjoint sets in free lattice ordered modules", *Houston J. Math.* (1989).

[1989b] W. B. Powell and C. Tsinakis, "The failure of the amalgamation property in varieties of representable ℓ-groups", *Math. Proc. Camb. Phil. Soc.* (1989).

[1989c] W. B. Powell and C. Tsinakis, "Covers of the variety of abelian ℓ-groups", *Comm. Alg.* (1989).

[1989d] W. B. Powell and C. Tsinakis, "Free products in varieties of ℓ-groups", *Lattice-Ordered Groups,* Reidel, 1989.

[1989e] W. B. Powell and C. Tsinakis, "Amalgamation in classes of lattice-ordered groups", *Lattice-Ordered Groups,* Reidel, 1989.

[1981] N. R. Reilly, "A subsemilattice of the lattice of varieties of lattice-ordered groups", *Can. J. Math.* 33(1981), 1309–1318.

[1983] N. R. Reilly, "Nilpotent, weakly abelian and Hamiltonian lattice ordered groups", *Czech. Math. J.* 33 (1983), 348–353.

[1986] N. R. Reilly, "Varieties of lattice-ordered groups that contain no non-abelian o-groups are solvable", *Order* 3(1986) 287–297.

[1989] N. R. Reilly, "Varieties of lattice-ordered groups", *Lattice-Ordered Groups,* Reidel, 1989.

[1975] E. B. Scrimger, "A large class of small varieties of lattice-ordered groups", *Proc. Amer. Math. Soc.* 51(1975), 301–306.

[1980] J. E. Smith, "The lattice of ℓ-group varieties", *Trans. Amer. Math. Soc.* 257 (1980), 347–357.

[1963] E. C. Weinberg, "Free lattice-ordered abelian groups", *Math. Ann.* 151(1963), 187–199.

[1965] E. D. Weinberg, "Free lattice-ordered abelian groups. II. ", *Math. Ann.* 159(1965), 217–222.

RECENT RESULTS ON THE FREE LATTICE-ORDERED GROUP OVER A RIGHT-ORDERABLE GROUP

Michael R. Darnel
Department of Mathematics
Indiana University at South Bend
South Bend, Indiana, USA

Abstract: In the following, we give a survey of recent results concerning when the free lattice-ordered group over a group retains some properties, such as nilpotency or solvability, of the original group.

Section One: Introduction Conrad [C] showed that for a partially ordered group G, a free lattice-ordered group over G exists if and only if the partial order of G is the intersection of right orders of G. If P is the cone of positive elements of a right order of G, then P^{-1} is also a cone of positive elements of a right order on G (by reversing the order) and the trivial order on G is the partial order obtained by intersecting P with P^{-1}. Thus it is easy to see that a free lattice-ordered group F(G) over the group G exists if and only if G can be right-ordered.

The construction of F(G) is easy to describe. Let $\{\leq_\lambda\}_\Lambda$ be the set of all right orders of G. For each $\leq_\lambda$, G acts (by multiplication on the right) as a group of order-preserving permutations of the chain $(G, \leq_\lambda)$. So by way of the right regular representation, G can be

J. Martinez (ed.), Ordered Algebraic Structures, 51–57.

embedded into $A(G, \leq_\lambda)$, the ℓ-group of all order-preserving permutations of the chain $(G, \leq_\lambda)$. $F(G)$ is then the ℓ-subgroup of $\Pi_{\lambda\varepsilon\Lambda} (G, \leq_\lambda)$ generated by the "long constants" of G: $g \to (\ldots, \bar{g}_\lambda, \ldots)$. More useful in the following discussion is that if G^*_λ is the ℓ-subgroup of $A(G, \leq_\lambda)$ generated by the right regular representation of G, then $F(G)$ is the ℓ-subgroup of $\Pi_\Lambda G^*_\lambda$ generated by the long constants of G. Thus if we can demonstrate that each G^*_λ is in a variety of lattice-ordered groups, then $F(G)$ must be as well.

Two final points need to be made. The first is that, unless $\leq_\lambda$ is a two-sided order, the embedding of G_λ into G^*_λ does not preserve the $\leq_\lambda$-ordering of G.

The second point is that G^*_λ is the sublattice of $A((G, \leq_\lambda)$ generated by (the right regular representation of) G. So [W] every element of G^* can be written in the (nonunique) form $\vee_I \wedge_J \bar{g}_{ij}$, where I and J are finite sets and each $\bar{g}_{ij}$ is an element of G [W]. Moreover, since in lattice-ordered groups, $(\vee_I \wedge_J g_{ij})(\vee_K \wedge_M h_{km}) = \vee_I \wedge_J \vee_K \wedge_M (g_{ij}h_{km})$, the free lattice-ordered group over a group G is abelian if and only if G is abelian. This result has recently been improved to the following:

<u>THEOREM 1</u>: [DGR] The free lattice-ordered group over a group G is representable (as a subdirect product of two-sided totally-ordered groups) if and only if G is torsion-free abelian.

A theorem much like Theorem 1 gives a characterization for when the free lattice-ordered group over G is normal-valued. Recall that a lattice-ordered group H is normal-valued if for any x, y ε H, $|x||y| \leq |y^2||x^2|$. This condition is easily converted to an equivalent equation and so the class of normal-valued lattice-ordered groups is a variety (or equationally defined class). Indeed, the variety of normal-valued lattice-ordered groups forms the largest proper variety of lattice-ordered groups and so any ℓ-group that is not normal-valued can not satisfy any ℓ-group equation not satisfied by all lattice-ordered groups. The defining law is equivalent to the condition $|[x, y]| \ll |x| \vee |y|$ and to the condition: if g, h $\ll$ k, then gh $\ll$ k. These three conditions are also equivalent for right-ordered groups, and a right-ordered group satisfying any (and hence all) of the conditions is called a Conrad right-ordered group, more commonly known as a c-group. The right order is then called a c-order.

THEOREM 2: [GHR] The free lattice-ordered group over a group G is normal-valued if and only if every right order of G is a c-order.

(What was actually shown in [GHR] that if (G, $\leq$) is a c-group, then G^* in (G,$\leq$) is normal-valued. Theorem 2 is then an easy consequence of this result.)

Section Two: The free lattice-ordered group over a solvable group

In the following, we will call a lattice-ordered group G ℓ-solvable of rank n if there exists a chain of convex ℓ-subgroups

$$(e) = A_0 \subset A_1 \subset \ldots \subset A_{n-1} \subset A_n = G$$

such that A_i is normal in A_{i+1} and each quotient A_{i+1}/A_i is abelian. Smith [S] pointed out that while ℓ-solvability of rank n implies solvability of rank n, the two concepts are not equivalent. In [GHM], it was shown that $\underline{wr^n}\ \mathbb{Z}$, the iterated ordered wreath product of the group $\mathbb{Z}$ of integers with itself n times, generates the variety of ℓ-solvable lattice-ordered groups of rank n.

Above we saw that the free lattice-ordered group over an abelian group is abelian. The natural question is: can we generalize 'abelian' to other properties such as solvable, metabelian, or nilpotent?

The answer to each is 'no'. To conclusively answer the first two, consider the free lattice-ordered group over the restricted wreath product G of $\mathbb{Z}$ with itself. This wreath product is of course a splitting extension of a countably infinite small sum of integers by the integers and so any element can be written uniquely in the form (f, m), where $f: \mathbb{Z} \to \mathbb{Z}$. As shown in Example 7.5.5 of [BMR], this wreath product can be right-ordered by declaring (f, m) to be positive if: (i) for k = max(supp(f)), f(k) > 0; or (ii) if f = 0, $m \geq 0$. But as also shown in [BMR], this right order is not a c-order. So embedding G into $A(G, \leq_r)$ by its right action yields that G^* is not normal-valued and so can not satisfy any law not satisfied by all lattice-ordered groups. So in particular, G^* is not solvable, let alone metabelian.

For nilpotency, we can get more satisfactory answers. In [DG] it was shown that the free lattice-ordered group over a finitely generated nilpotent group of class two must be ℓ-solvable and moreover that the free lattice-ordered group over the free nil-2 group of rank two actually generates the variety of lattice-ordered groups that are ℓ-solvable of rank two. These results have been generalized to all finitely generated torsion-free nilpotent groups.

To begin, Rhemtulla [R] proved that any right order of a nilpotent group must be a c-order. Thus the free lattice-ordered group over a torsion-free nilpotent group must be normal-valued.

A key lemma is the following:

LEMMA 3: [D1] Let $(G, \leq_r)$ be a right-ordered group and let K be the convex subgroup of G generated by the derived subgroup $G^{(1)}$. If $K \neq G$, then G^* is a lex extension of K^*.

In particular, if $(G, \leq_r)$ is a finitely generated c-group, it can be shown that at least one generator must be infinitely greater than any commutator and so the above lemma applies.

Recall that the Hirsch length of a finitely generated torsion-free nilpotent group is the length of any central series

$$(e) = A_0 \lhd A_1 \lhd \dots \lhd A_n = G$$

such that A_{i+1}/A_i is free abelian of rank one.

THEOREM 4: [D1] Let G be a finitely generated torsion-free nilpotent group of Hirsch length n. Then the free lattice-ordered group F(G) over G is ℓ-solvable of rank at most n.

About all that can be said for the above bound for the ℓ-solvability rank is that it is clear. The bound can be made much smaller if we restrict our attention to the torsion-free nilpotent groups of class two.

THEOREM 5: [D1] Let G be a finitely generated torsion-free nil-2 group and let n be the minimal number of generators necessary to generate G. Then the free lattice-ordered group F(G) over G is ℓ-solvable of rank at most n.

This last theorem is the best possible. If G_n is the free nil-2 group on n generators, it is shown in [D1] that a c-order $\leq_r$ can be placed upon G_n such that the lattice-ordered iterated wreath product $wr^n \mathbb{Z}$ can be ℓ-embedded into G_n^* in $A(G_n, \leq_r)$. Thus the free lattice-ordered group $F(G_n)$ over G_n actually generates the ℓ-solvable variety of rank n. This incidentally shows that $F(G_n)$ is not nilpotent.

Finally, let A be a torsion-free abelian group. Define the A-Scrimger n-group, denoted $S_n(A)$, to be the splitting extension of $\sum_{i=1}^{n} A$ with the group $\mathbb{Z}$ of integers where

$$-1_{\mathbb{Z}} + (k_1, k_2, \ldots, k_n) + 1_{\mathbb{Z}} = (k_n, k_1, k_2, \ldots, k_{n-1}).$$

It is readily verified that for any torsion-free abelian group A and any positive integer n, $S_n(A)$ satisfies the commuting n^{th}-power law: for all x and y, $x^n y^n = y^n x^n$.

THEOREM 6: [D2] For any torsion-free abelian group A and any positive integer n, the free lattice-ordered group over $S_n(A)$ satisfies the commuting n^{th}-power law.

REFERENCES:

[BMR] Botto Mura, R.; and Rhemtulla, A.; "Orderable Groups," Marcel Dekker, 1977.

[C] Conrad, P.F.; "Free lattice-ordered groups," J. Algebra, 16(1970), 191-203.

[D1] Darnel, M.R., "The free lattice-ordered group over a nilpotent group," submitted to Proceedings of the Amer. Math. Soc.

[D2] Darnel, M.R., "The free lattice-ordered group over a commuting n^{th}-powers group," in preparation.

[DG] Darnel, Michael and Glass, A.M.W.; "Commutator relations and identities in lattice-ordered groups," accepted by Michigan Math. J.

[DGR] Darnel, M.R.; Glass, A.M.W.; and Rhemtulla, A.; "Groups in which every right order is two sided," accepted by Archiv der Matematik.

[GHR] Glass, A.M.W.; Hollister, H.; and Rhemtulla, A.; "Right orderings versus lattice-orderings," Notices of the AMS, 2(1978), A-222.

[R] Rhemtulla, A.; "Right-ordered groups," Canad. J. Math., 24(1972), 891-895.

[S] Smith, J.E.; "Solvable and ℓ-solvable ℓ-groups," Algebra Universalis, 18(1984), 106-109.

[W] Weinberg, E.; "Free lattice-ordered abelian groups," Math. Ann., 151(1963), 187-199.

NON-AMALGAMATION AND CONJUGATION FOR ℓ-METABELIAN LATTICE-ORDERED GROUPS

A. M. W. Glass[1]
Department of Mathematics and Statistics
Bowling Green State University
Bowling Green, Ohio 43403-0221 USA

To L. I. Glass--in memoriam.

ABSTRACT. In this simple paper (which is an off-shoot of far more difficult work attempted with J. S. Wilson) we consider amalgamation and conjugation questions for ℓ-metabelian lattice-ordered groups. The first section gives a trivial example to show that one of the few positive amalgamation results cannot be generalised; the second begins a study of conjugation in subclasses of lattice-ordered groups (as opposed to the class of all lattice-ordered groups).

1. Amalgamation

Let $\mathcal{C}$ be a class of ℓ-groups and $G, H_1, H_2 \in \mathcal{C}$, and let $\sigma_i : G \to H_i$ be ℓ-group embeddings for $i = 1,2$; then $(G, H_1, H_2, \sigma_1, \sigma_2)$ is called a *V-formation in* $\mathcal{C}$ *with base* G. We say that the V-formation has an *amalgam in* $\mathcal{C}$ if there exist an ℓ-group $K \in \mathcal{C}$ and ℓ-group embeddings $\tau_i : H_i \to K$ for $i = 1,2$ such that $\sigma_1\tau_1 = \sigma_2\tau_2$. If every V-formation in $\mathcal{C}$ has an amalgam in $\mathcal{C}$, then $\mathcal{C}$ is said to have the *amalgamation property*.

In [8], K. R. Pierce proved the following:

THEOREM 1: *If* $\mathcal{C}$ *is the class of all Abelian ℓ-groups or the class of all Abelian o-groups, then* $\mathcal{C}$ *has the amalgamation property.*

In [7] the same author proved that if $\mathcal{C}$ is the class of all ℓ-groups, then any V-formation with an Archimedean o-group as base has an amalgam in $\mathcal{C}$. In [10] Reilly showed using permutational products that if $\mathcal{C}$ is the class of all ℓ-groups or the class of all o-groups, then any V-formation in which $G\sigma_i$ is a convex ℓ-subgroup

[1]Research supported by an NSF US-UK grant. I am extremely grateful to the NSF for making this research possible and to members of the Department of Pure Mathematics and Mathematical Statistics of the University of Cambridge for their hospitality and facilities. I am especially grateful to John Wilson for very many helpful discussions and for finding an error in an earlier draught of this paper.

J. Martinez (ed.), Ordered Algebraic Structures, 59–64.

of H_i contained in its centre for $i = 1,2$ has an amalgam in $\mathcal{C}$. By a result of Holland (see [3, Theorem 11F]), Reilly's proof actually yields the corresponding result with $\mathcal{C}$ any variety of ℓ-groups or the class of all o-groups in any given variety of ℓ-groups.

On the negative side, the class of all ℓ-groups fails to have the amalgamation property [7] as do the class of all o-groups [4] and even the class of totally orderable groups [5, page 30]. Indeed, the amalgamation property fails quite badly for ℓ-metabelian ℓ-groups. (We recall that an ℓ-group H is called *ℓ-metabelian* if it has an Abelian ℓ-ideal A such that H/A is Abelian.) In [7], it is shown that V-formations with G and H_1 Abelian and H_2 ℓ-metabelian need not have an amalgam; and in [4] and [9] the failure is exhibited when G, H_1 and H_2 are ℓ-metabelian o-groups. In the examples in these three papers, one even has $G\sigma_i < H_i$ for $i = 1,2$ though $G\sigma_i$ is not convex in H_i $(i = 1,2)$.

In this section, we wish to show that Reilly's result is tight for ℓ-metabelian ℓ-groups. Specifically

THEOREM 2. *Let $\mathcal{C}$ be the class of all ℓ-metabelian ℓ-groups. Then there is a V-formation $(G,H_1,H_2,\sigma_1,\sigma_2)$ in $\mathcal{C}$ with G Abelian and $G\sigma_i$ an ℓ-ideal of the ℓ-metabelian o-group H_i for $i = 1,2$, such that no amalgam is possible in $\mathcal{C}$.*

This result contrasts with Bludov and Medvedev's theorem that every ℓ-metabelian o-group can be embedded as an ℓ-group in a divisible such--see [1].

<u>*Proof of Theorem 2:*</u> Let F be the free group on two free generators α_1 and α_2. By [2, Section IV.2], F can be made into an o-group in such a way that $\alpha_i > e$ and the convex subgroup of F generated by α_i is F for $i = 1,2$. Let $\mathbb{Z}$ denote the additive

o-group of integers and $G = \overleftarrow{\sum_{u\in F}} \mathbb{Z}$, an Abelian o-group. Then α_i induces an o-automorphism of G in the natural way. Hence $H_i = G \quad \langle\alpha_i\rangle$ is an ℓ-metabelian o-group and if $\sigma_i : G \to H_i$ is the natural embedding, then $G\sigma_i$ is an ℓ-ideal of H_i for $i = 1,2$. Let $w = w(\alpha_1,\alpha_2)$ be any non-identity element of F. In any group amalgam of the V-formation $\langle G,H_1,H_2,\sigma_1,\sigma_2\rangle$, $w(\alpha_1\tau_1,\alpha_2\tau_2)$ has a non-identity action on $G\sigma_1\tau_1 = G\sigma_2\tau_2$. To see this, suppose

$w(\alpha_1,\alpha_2) = \alpha_1^{m_{11}}\alpha_2^{m_{12}}\cdots\alpha_1^{m_{n1}}\alpha_2^{m_{n2}}$ for some integers m_{ji} $(i = 1,2;$ $j = 1,\ldots,n)$ with $m_{12},\ldots,m_{n1} \neq 0$. Let $g_u = \begin{cases} 1 & \text{if } u = e \\ 0 & \text{if } u \neq e \end{cases}$; then

$(g\sigma_1\tau_1)^{w(\alpha_1\tau_1,\alpha_2\tau_2)} = f\sigma_1\tau_1$ where $f_u = \begin{cases} 1 & \text{if } u = w(\alpha_1,\alpha_2) \\ 0 & \text{otherwise} \end{cases}$ (and

$f\sigma_1\tau_1 \neq g\sigma_1\tau_1$ since $w(\alpha_1,\alpha_2) \neq e$). Hence the V-formation cannot have an amalgam in any variety of ℓ-groups that satisfies a non-trivial group identity. Since any ℓ-metabelian ℓ-group satisfies the group identity $[[x_1,x_2],[x_3,x_4]] = e$, the theorem follows. //

There are two non-nilpotent varieties of ℓ-groups that cover the variety of all Abelian ℓ-groups and are generated by ℓ-metabelian o-groups. These are denoted by $\mathcal{M}^+$ and $\mathcal{M}^-$ (see [6]). As in [9],

$H_i \in \mathcal{M}^+$ for $i = 1,2$. Replacing G by $\vec{\sum_{u \in F}} \mathbb{Z}$ in the above, we can obtain $H_i \in \mathcal{M}^-$ $(i = 1,2)$. So the proof of Theorem 2 actually yields

THEOREM 3. *If C is any class of ℓ-groups that contains $\mathcal{M}^+$ (or $\mathcal{M}^-$) and satisfies some non-trivial group identity, then there is a V-formation $(G,H_1,H_2,\sigma_1,\sigma_2)$ in C with G Abelian, $H_i \in \mathcal{M}^+$ (or $\mathcal{M}^-$) an o-group and $G\sigma_i$ and ℓ-ideal of H_i, for $i = 1,2$, which has no amalgam in C.*

2. Conjugation

Conjugation and amalgamation are closely related. In [7], K. R. Pierce proved that if G is an ℓ-group and $e < f,g \in G$, there is an ℓ-group H and an ℓ-embedding φ of G into H such that $f\varphi$ and $g\varphi$ are conjugate in H. From this he deduced that any V-formation in the class of all ℓ-groups with base $\mathbb{Z}$ has an amalgam in the class of all ℓ-groups. The ℓ-group H in his proof generates the variety of all ℓ-groups no matter what the ℓ-group $G \neq \{e\}$ is.

We wish to examine the problem when G and H are restricted to particular classes of ℓ-groups.

In this paper, the first to examine this natural question, we will be quite modest and restrict G to be an Abelian o-group or an Abelian ℓ-group and confine H to be an ℓ-metabelian o-group, a representable ℓ-metabelian ℓ-group, or an ℓ-metabelian ℓ-group.

THEOREM 4: *Let G be an Abelian o-group and f and g be strictly positive elements of G. Then G can be o-embedded in an ℓ-metabelian o-group H in which the images of f and g are conjugate.*

Proof: Any Abelian o-group can be o-embedded in an ordered field (see, e.g., [3, page 74]); so we may assume that G itself is an ordered field. Let θ be the automorphism of the Abelian o-group G given by multiplication by g/f. Then $H = G \rtimes \langle\theta\rangle$ is the desired ℓ-metabelian o-group. //

If we allow G to be an Abelian ℓ-group, is the corresponding result true for some ℓ-metabelian ℓ-group H? Since every Abelian ℓ-group is a subdirect product of Abelian o-groups, the strongest result one could hope for is that H could be taken to be an ℓ-metabelian representable ℓ-group. This is false:

EXAMPLE: The crux of the proof is that in any representable ℓ-group, if $a \wedge b = e$ then $h^{-1}ah \wedge b = e$.

Let $G = \mathbb{Z} + \mathbb{Z}$ and $k = (1,-1)$. Then $f = k^+ = (1,0)$, and $g = (1,1) = |k|$ are strictly positive elements of G and $k^- \neq e$. Moreover, $k^+ \wedge k^- = e$ and $k^- = |k| \wedge k^-$. Hence if H is any representable ℓ-group and φ is any ℓ-embedding of G into H, $|k|\varphi \wedge k^-\varphi = k^-\varphi \neq e$ and $k^+\varphi \wedge k^-\varphi = e$. Consequently, there is no element of H conjugating $f\varphi = k^+\varphi$ to $g\varphi = |k|\varphi$.

A somewhat weaker conclusion is true, however.

THEOREM 5: *Let G be an Abelian ℓ-group and f and g be strictly positive elements of G. Then G can be ℓ-embedded in an ℓ-metabelian ℓ-group in which the images of f and g are conjugate.*

Proof: Since any Abelian ℓ-group is a subdirect product of Abelian o-groups, we may assume that G is the full cardinal Cartesian power of some Abelian o-group. As in the proof of Theorem 4, this Abelian o-group can be taken to be an ordered field, say F. So $G = \Pi\{F: \lambda \in \Lambda\}$ for some index set Λ. If we can find an ℓ-group automorphism θ of an Abelian ℓ-group $\overline{G}$ (containing G as an ℓ-subgroup) mapping f to g, then the splitting extension $\overline{G} \rtimes \langle\theta\rangle$ will be the desired ℓ-metabelian ℓ-group. This we now proceed to obtain.

First, we obtain a set $M \supseteq \Lambda$ and an ℓ-embedding $\varphi: G \to \overline{G} = \Pi\{F: \mu \in M\}$ such that for some $\psi^* \in Sym(M)$, $supp(f\varphi)\psi^* = supp(g\varphi)$. Without loss of generality, we may assume that $|supp(f) \cup supp(g)| < |\Lambda|$.

Let $\Lambda_1 = supp(f)\backslash supp(g)$, $\Lambda_2 = supp(g)\backslash supp(f)$ and $\Lambda_3 = supp(f) \cap supp(g)$.

Case 1 $\Lambda_1 = \Lambda_2 = \emptyset$. Let $M = \Lambda$, φ be the identity ℓ-embedding and ψ^* be the identity permutation of M.

Case 2 $\Lambda_1 \neq \emptyset$ and $\Lambda_2 \neq \emptyset$. If $|\Lambda_1| = |\Lambda_2|$, let $M = \Lambda$, φ be the identity ℓ-embedding and $\psi^* \in Sym(M)$ be such that $\Lambda_1\psi^* = \Lambda_2$ and $\Lambda_3\psi^* = \Lambda_3$. If $|\Lambda_1| \neq |\Lambda_2|$, say $|\Lambda_1| < |\Lambda_2|$, let $\lambda_1 \in \Lambda_1$ and let M_1 be a set whose intersection with Λ is $\{\lambda_1\}$ with $|\Lambda_1 \cup M_1| = |\Lambda_2|$. Let $M = \Lambda \cup M_1$ and $\varphi: G \to \overline{G}$ be given by

$$(a\varphi)_\mu = \begin{cases} a_\mu & \text{if} \quad \mu \in \Lambda \\ a_{\lambda_1} & \text{if} \quad \mu \in M_1 \end{cases}.$$

Then φ is an ℓ-embedding and there is $\psi^* \in Sym(M)$ such that $(\Lambda_1 \cup M_1)\psi^* = \Lambda_2$ and $\Lambda_3\psi^* = \Lambda_3$. Then $supp(f\varphi)\psi^* = (\Lambda_1 \cup M_1 \cup \Lambda_3)\psi^* = \Lambda_2 \cup \Lambda_3 = supp(g\varphi)$ as desired. If $|\Lambda_2| < |\Lambda_1|$ we proceed similarly.

Case 3 $\Lambda_1 = \emptyset$ and $\Lambda_2 \neq \emptyset$. Since $supp(f) = \Lambda_3$ and $f > 0$, $\Lambda_3 \neq \emptyset$. Let $\lambda_3 \in \Lambda_3$ and let M_3 be a set of cardinality $\max\{\aleph_0, |\Lambda_2|\}$ whose intersection with Λ is $\{\lambda_3\}$. Let $M = \Lambda \cup M_3$

and $\varphi: G \to \overline{G}$ be given by $(a\varphi)_\mu = \begin{cases} a_\mu & \text{if } \mu \in \Lambda \\ a_{\lambda_3} & \text{if } \mu \in M_3 \end{cases}$. Then φ is an ℓ-embedding and there is $\psi^* \in \mathrm{Sym}(M)$ such that $M_3\psi^* = \Lambda_2 \cup M_3$ and $\Lambda_3\psi^* = \Lambda_3$. Then $\mathrm{supp}(f\varphi)\psi^* = (\Lambda_3 \cup M_3)\psi^* = \Lambda_3 \cup \Lambda_2 \cup M_3 = \mathrm{supp}(g\varphi)$.

Case 4 $\Lambda_1 \neq \varnothing$ and $\Lambda_2 = \varnothing$. Proceed similarly to case 3.

Now ψ^* induces an ℓ-automorphism of $\overline{G}$, say ψ, in the natural way: $(a\psi)_\mu = a_{\mu\psi^*}$ $(\mu \in M)$. Moreover, $\mathrm{supp}(f\varphi\psi) = \mathrm{supp}(g\varphi)$. Let η be the ℓ-automorphism of $\overline{G}$ obtained by multiplying the μth coordinate by 1 if $\mu \notin \mathrm{supp}(f\varphi\psi)$ and by $(g\varphi\psi)_\mu/(f\varphi\psi)_\mu = (g\varphi)_{\mu\psi^*}/(f\varphi)_{\mu\psi^*}$ if $\mu \in \mathrm{supp}(f\varphi\psi)$. Then $(f\varphi)\psi\eta = g\varphi$, so $\psi\eta$ is the desired ℓ-automorphism of $\overline{G}$. Thus the theorem is proved. //

Obviously there are many other classes to consider. As the example above shows, no such result is possible for G and H subdirect product of o-groups. However, the question remains open for $G \in \mathcal{A}^n$ and $H \in \mathcal{A}^{n+1}$ if $n > 1$ (with f and g appropriately restricted), where $\mathcal{A}$ is the variety of all Abelian ℓ-groups. Even more naturally, the case that G and H belong to the largest proper variety of ℓ-groups, the class of all normal-valued ℓ-groups, needs to be explored. These problems are being examined by Phil Moore in his Ph.D. thesis.

References

1. V. V. Bludov and N. Ya Medvedev, The completion of orderable metabelian groups, *Algebra i Logika*, **13**(1974), 369-373.

2. L. Fuchs, *Partially ordered algebraic systems*, Pergamon Press, 1963.

3. A. M. W. Glass, *Ordered Permutation Groups*, London Math. Soc. Lecture Notes, Series No. 55, Cambridge University Press, 1981.

4. A. M. W. Glass, D. Saracino and C. Wood, Non-amalgamation of ordered groups, *Math. Proc. Cambridge Phil. Soc.* **95**(1984), 191-195.

5. A. I. Kokorin and V. M. Kopytov, *Fully ordered groups*, John Wiley & Sons, New York, 1974.

6. N. Ya Medvedev, Lattices of varieties of lattice-ordered groups and Lie algebras, *Algebra and Logic* **16**(1977), 27-30.

7. K. R. Pierce, Amalgamations of lattice-ordered groups, *Trans. Amer. Math. Soc.* **172**(1972), 249-260.

8. K. R. Pierce, Amalgamating Abelian ordered groups, *Pacific J. Math* 43(1972), 711-723; amalgamated sums of Abelian ℓ-groups, *Pacific J. Math* 65(1976), 167-173.

9. W. B. Powell and C. Tsinakis, The failure of the amalgamation property for representable varieties of lattice-ordered groups, *Math. Proc. Cambridge Phil. Soc.* (to appear).

10. N. R. Reilly, Permutational products of lattice-ordered groups, *J. Australian Math. Soc.* 13(1972), 25-34.

PERIODIC EXTENSIONS OF ORDERED GROUPS

A.H. Rhemtulla
Department of Mathematics
University of Alberta
Edmonton, Canada T6G 2G1

1. Introduction

A PARTIAL ORDER P on a group G is a subset P of G satisfying the following conditions:

(i) $PP = 1$

(ii) $P \cap P^{-1} = 1$ and

(iii) $Pg = gP$ for all g in G.

In the usual terminology of partial order relation $\leq$, this is equivalent to saying $a \leq b$ if and only if $ba^{-1} \in P$. A partial order P is a total order if it also satisfies condition

(iv) $P \cup P^{-1} = G$.

Clearly every partial order P can be extended to a maximal partial order on G, and an old and natural problem is the study of groups in which every partial order can be extended to a total order. This class, named 0^*-groups by Fuchs [2] has been extensively studied. It includes all torsion-free nilpotent groups, all ordered solvable groups of finite Prüfer rank and all ordered metabelian groups. But a centre by metabelian ordered group need not be in this class.

Other related questions such as extensions of partial orders to directed orders or to isolated partial orders or to lattice orders (Questions 2, 3, 4 of [2]) have received less attention. Details are contained in [2], [3] and [6]. Others, including B.H. Neumann and J.A.H. Sheppard, have considered the problem of extension of certain types of partial orders to total orders. Let H be a normal subgroup of a torsion-free group G. In [7] Neumann and Sheppard have proved that if G/H is locally finite then any G-order on H can be extended to an order on G . They ask if the result would hold if the restriction on G/H is weakened from locally finite to periodic. S.I. Adian (see [1] Section 7.1) has constructed groups $A = A(m,n)$ for any $m > 1$ and any odd $n \geq 665$ which are torsion-free, A/Z is the free Burnside group on m generators and exponent

J. Martinez (ed.), Ordered Algebraic Structures, 65–69.

n and $Z = Z(A)$, the centre of A, is cyclic. These groups provide counter examples for all odd $n \geq 665$. When n is a power of 2 the question remains open.

The main results are the following.

THEOREM 1. *If H is a locally nilpotent normal subgroup of a torsion free group G and G/H is a periodic 2-group or a periodic Engel group then any G-order on H can be extended to an order on G. In both cases every partial order on H can be extended to a (total) order on G.*

THEOREM 2. *Let P be a G-order on a normal subgroup H of G such that the chain of convex subgroups of H under P is ascendent. If G is torsion-free and G/H is a periodic 2-group or a periodic Engel group then P can be extended to an order on G. In particular if H is solvable of finite Prüfer rank then any G-order on H can be extended to an order on G.*

For any elements x, y of a group G we write $[x, y]$ to denote $x^{-1}y^{-1}xy$. If $r \geq 0$ is an integer then $[x, ry]$ will denote $[x, \underbrace{y, \ldots, y}_{r}]$ which is defined to be x if $r = 0$ and $[[x, (r-1)y], y]$ otherwise. G is an Engel group if for each pair x, y of elements of G there exists an integer $r = r(x, y) \geq 0$ such that $[x, ry] = 1$. A group G is said to have finite Prüfer rank n if every finitely generated subgroup of G can be generated by n elements. For further terminology and definitions of terms used we refer the reader to [4] for group theoretic terms and [6] for terms dealing with order structure.

If one does not impose any restriction on the subgroup H, then G may as well be taken to be a free group. For if H is a normal subgroup of a torsion-free group G such that G/H lies in some given class $\mathcal{X}$ of groups, and P is a G-order on H, then let F be an appropriate free group such that G is isomorphic to F/K for some normal subgroup K of F. Let $\mathcal{O} : G \to F/K$ denote this isomorphism and let $L/K = \mathcal{O}(H)$. Since F is a free group, it is orderable. Let Q be any order on F so that $Q \cap K$ is an F-order on K. Extend this order to an F-order Q_1 on L by putting any $g \in L \backslash K$ in Q_1 if $gK \in \mathcal{O}(P)$. Now $F/L \in \mathcal{X}$ and Q_1 is an F-order on L. Moreover Q_1 can be extended to an order on F if and only if P can be extended to an order on G. This shows that if no assumptions are made regarding the subgroup H, then the additional hypothesis that G is orderable does not matter in deciding whether the G-order P on H can be extended to an order on G.

In view of this, the portion of Neumann and Shepperd's question that still remains open can be rephrased as follows.

Question 1. Let H be a normal subgroup of a free group F such that F/H is a periodic 2-group. Can every F-order on H be extended to an order on

F ?

The following question is also open.

Question 2. Let H be a normal subgroup of a free group F such that F/H is a periodic Engel group. Can every F-order on H be extended to an order on F ?

Even if the questions above have a negative answer, the following may have affirmative solutions.

Question 3. Let H be a solvable normal subgroup of a torsion-free group G such that G/H is a periodic 2-group. Can every G-order on H be extended to an order on G ?

Question 4. Let H be a solvable normal subgroup of a torsion-free group G such that G/H is a periodic Engel group. Can every G-order on H be extended to an order on G ?

2. Proofs

We outline the proof of Theorem 1. The proof of the second result is similar. Let P be a G-order on H and let

$$Q = \{g \in G : \ g^n \in P \quad \text{for some} \quad n > 0\}.$$

Since P is a G-order, Q is G-invariant. Since G is torsion-free, $Q \cap Q^{-1} = < e >$ and since G/H is periodic $Q \cup Q^{-1} = G$. Thus it remains to show that Q is a semigroup. Pick any x, y in Q . In order to show that $xy \in Q$ it is sufficient to assume that $G = < x, y >$.

If $H \leq L \triangleleft G$ and $P(L) = \{g \in L; \ g^n \in P \ \text{ for some } \ n > 0\}$ is a semigroup, then $P(L)$ is a G-order on L . Since L is orderable, L/H is periodic and H is locally nilpotent then it follows that L is locally nilpotent (see [6], Corollary 2.6.7). Thus the hypotheses of the theorem remain valid if H is replaced by L and P is replaced by $P(L)$. Similarly if $H \leq L_1 \leq L_2 \leq \dots$ is an ascending sequence of normal subgroups of G , $L = \bigcup_{i=1}^{\infty} L_i$ and $P(L_i) = \{g \in L_i; g^n \in P \ \text{ for some } \ n > 0\}$ is a semigroup for all $i = 1, 2, \dots$ then $\bigcup_{i=1}^{\infty} P(L_i) = P(L)$ is a semigroup and $P(L)$ is a G-order on L . Thus we may assume that H is maximal in the sense that if $H < L \triangleleft G$ then $P(L)$ is not a semigroup.

Since G/H is periodic, $K = \langle x^n, y^n \rangle \leq H$ for some $n \neq 0$. If G/H is a 2-group then we may assume $n = 2^m$ for some $m \geq 0$. Now K is nilpotent. Let $\zeta_i(K)$ denote the i-th term of the upper central series of K and let Z_i denote the i-th term of the upper central series of G . We show, via induction on i , that $\zeta_i(H) = Z_i \geq \zeta_i(K)$ and Z_i is an isolated subgroup of G. For $i = 0, <e> = Z_0 = \zeta_0(K)$ and since G is torsion-free, $<e>$ is isolated. So assume the result holds for some $i \geq 0$ and let $\zeta = \zeta_{i+1}(K)$. Since $<H, x>$ is a finite extension of H, it is an orderable group by the principal result of [7]. Hence $<H, x>/Z_i$ is an orderable group (see [3], p. 26 or [6] Theorem 2.2.4). Now pick any $a \in \zeta$. Then $[a, x^n] \in Z_i$. Hence $[a,x][a,x]^x \dots [a,x]^{x^{n-1}} \in Z_i$. Thus $[a,x] \in Z_i$. Similarly $[a,y] \in Z_i$ and hence $[a, G] \leq Z_i$ or $a \in Z_{i+1}$. Now suppose that for some $h \in H$, $h^r \in Z_{i+1}$ for some $r > 0$. Then $[x, h^r] \in Z_i$. Hence $[x,h][x,h]^h \dots [x,h]^{h^{r-1}} \in Z_i$ and since $[x,h] \in H$, it follows that $[x,h] \in Z_i$. Similarly $[y,h] \in Z_i$ and hence $h \in Z_{i+1}$.

Finally suppose $g \in G$ and $g^k \in Z_{i+1}$ for some $k > 0$. For any $h \in H$, $[g^k, h] \in Z_i$ implies $[g,h] \in Z_i$ since $\langle H, g\rangle$ is a finite extension of H, hence orderable and hence $\langle H, g\rangle/Z_i$ is orderable. So we need to show that $[g,x]$ and $[g,y]$ are in Z_i. Notice that up to this point we have only used the fact that G/H is periodic.

Suppose G/H is a 2-group then $x^{2^m} \in H$ for some $m \geq 0$ and hence $[g, x^{2^m}] \in Z_i$. It is sufficient to assume that $[g, x^2] \in Z_i$ and show $[g,x] \in Z_i$. Since $[g,x^2] = [g,x][g,x]^x$, $[g,x]^x \equiv [g,x]^{-1}$ Mod Z_i. But $[g,x]^n \in H$ for some n. Hence $([g,x]^n)^x \equiv [g,x]^{-n}$ Mod Z_i. But $\langle H, x\rangle$ is orderable and so is $\langle H,x\rangle/Z_i$. Thus $[g,x]^n \equiv 1$ Mod Z_i. Since Z_i is isolated in G, $[g,x] \in Z_i$, as required.

Suppose G/H is a periodic Engel group. Then $[x, rg] \in H$ for some $r > 0$. Let ℓ be the least integer such that $[x\ \ell g] \in Z_i$. This will happen for some integer ℓ as can be seen as follows. Since G/H is an Engel group, $[x, rg] \in H$ for some $r \geq 0$. Since $g^k \in Z_{i+1}, [x, rg, g^k] = [x,(r+1)g]$ $[x,(r+1)g]^g \dots [x,(r+1)g]^{g^{k-1}} \in Z_i$ with $[x,(r+1)g] \in H$. But $\langle H, g\rangle$ and hence $\langle H,g\rangle/Z_i$ is orderable. Thus $[x,(r+1)g] \in Z_i$. If $\ell \neq 1$ then $[x,(\ell-2)g, g^k] \equiv [x,(\ell-1)g]^k$ Mod Z_i. Hence $[x,(\ell-1)g] \in Z_i$, a contradiction. Thus $[x,g] \in Z_i$ as required.

We have now shown that Z_{i+1} is isolated in both cases. By maximality of H it is clear that $Z_{i+1} \subseteq H$ since $Z_{i+1}H/H$ is locally finite and hence any G-order on H can be lifted to a G-order on $Z_{i+1}H$ by the main result of [7]. If $h \in \zeta_{i+1}(H)$ then $[h, x^n] \in Z_i$. But $\langle H,x\rangle/Z_i$ is orderable. Hence $[h,x] \in Z_i$ and $h \in Z_{i+1}$. Thus $\zeta_{i+1}(H) \subseteq Z_{i+1}$ and hence $\zeta_{i+1}(H) = Z_{i+1}$.

Since K is nilpotent, $\zeta_c(K) = K$. Hence $Z_c = G$ and G is nilpotent. By our choice of H we have $H = G$. That every partial order on H can be extended to an order on G now follows from Malćev's result in [5].

3. Finite extensions of right orders

It is natural to ask if the Neumann-Sheppard result has an analogue for right orders. The following example shows that not every right order on a subgroup of index two in G can be extended to a right order on G where G is a metabelian right ordered group. However the result is true for locally nilpotent-by-finite groups as in this case every partial right order can be extended to a total right order (Theorem 7.6.4 [6]).

Let Q denote the additive group of all rational numbers of the form $m/2^n$, $m, n \in \mathbb{Z}$. Let $G = \langle Q, t\rangle$ be the split extension of Q by an infinite cyclic group $\langle t\rangle$ where $t^{-1}qt = -2q$ for all q in Q. Thus elements of G have the form (q, t^n) and $(q_1, t^r)(q_2, t^s) = (q_1 + q_2/(-2)^r, t^{r+s})$.

Let $H = \langle Q, t^2\rangle$ so that H is of index two in G. Define subset $P \subseteq H$ by the rule: $(q, t^{2n}) \in P$ if $q > 0$ or $q = 0$ and $n \geq 0$. It is easy to see that $PP = P, P \cap P^{-1} = \{0\}$ and $P \cup P^{-1} = H$ so that P is a right order on H. Suppose that P can be extended to a right order P_1 on G. Then $t \in P_1$ since $t^2 \in P$ and any $q > 0$, $q \in Q$ is also in P_1. Hence P_1 contains $(0,t)(q,1)$. But $(0,t)(q,1)(0,t)(q,1) = (-q/2 + q/4, t^2) \in P^{-1}$ since $-q/2 + q/4 < 0$, a contradiction. Thus P can not be extended to a right order on G.

References

[1] Adian, S.I., *The Burnside Problem and Identities in Groups.* Translated from the Russian by John Lennox and James Wiegold. Ergebnisse der Mathematik und ihrer Grenzgebiete, Band 95. Springer-Verlag New York, Inc., New York, 1979.

[2] Fuchs, L., *Partially Ordered Algebraic Systems.* International Series of Monographs on Pure and Applied Mathematics, vol. 28, Pergamon Press (1963).

[3] Kokorin, A.I. and Kopytov, V.M., *Fully Ordered Groups.* Translated from Russian by D. Louish. John Wiley & Sons, New York 1974.

[4] Kurosh, A.G., *Theory of Groups*, Volume Two. Translated from Russian by K.A. Hirsch. Chelsea Publishing Company, New York, 1960.

[5] Malćev, A.I., *On the Full Ordering of Groups.* Trudy Mat. Inst. Stiklov. **38**, 173–175 (1951).

[6] Mura, R.B. and Rhemtulla, A., *Orderable Groups*, Dekker, New York 1977.

[7] Neumann, B.H. and Shepperd, J.A.H., *Finite Extensions of Fully Ordered Groups.* Proc. Royal Soc. London, Ser A, **239**, 320–327 (1957).

THE GATE COMPLETION $G^{\because}$ OF A LATTICE-ORDERED PERMUTATION GROUP

Richard N. Ball
University of Denver
Denver, Colorado 80208

and

Stephen H. McCleary
University of Georgia
Athens, Georgia 30602

The *gate completion* $G^{\because}$ of an ℓ-permutation group (G,Ω) is the ℓ-subgroup of $A(\bar{\Omega})$ consisting of those elements which "respect the tyings" (roughly, the equality of stabilizer subgroups) of (G,Ω), or equivalently, which can be "finitely gated" by elements of G (detailed definitions later). We assume throughout that the dead segments of (G,Ω) have been collapsed to points. When (G,Ω) has (order) closed stabilizer subgroups (and every completely distributive ℓ-group can be so represented), the ℓ-group $G^{\because}$ turns out to be independent of the representation, being in fact the α-completion G^{α} of G.

We explore the question of how $G^{\because}$ can be built up from G by adjoining infinite suprema and infima of subsets of G, and iterating. Because $(G^{\because},\bar{\Omega})$ has closed stabilizer subgroups (regardless of whether (G,Ω) does), these suprema and infima refer equivalently to either $G^{\because}$ or $A(\bar{\Omega})$. Not only does the order closure of G in either turn out to be $G^{\because}$, but each $h \in G^{\because}$ can be written as $h = \bigvee_i \bigwedge_j \bigvee_k g_{ijk}$ ($g_{ijk} \in G$). Can we do even better, writing $h = \bigvee_i \bigwedge_j g_{ij}$ ($g_{ij} \in G$)? In general, no, but this *is* true when Ω is countable! It is this last result that we shall prove here, making the most intense use ever of the structure theory of ℓ-permutation groups (but for simplicity

J. Martinez (ed.), Ordered Algebraic Structures, 71–92.

assuming that (G,Ω) has convex orbits, which will mean that only the transitive structure theory is actually needed). When G is completely distributive, our results elucidate the relationship of G^α to G, and in particular establish that G is dense in the α-topology on G^α.

Background on ℓ-permutation groups can be found in [G].

1. Definitions

A *dead segment* of (G,Ω) is a nonsingleton o-block Δ such that $G_\Delta|\Delta = \{e\}$ (where $G_\Delta = \{g \in G|\ \Delta g = \Delta\}$); and such that $\{\Delta g|\ g \in G,\ \Delta < \Delta g\}$ has no smallest element and no point of Ω lies between it and Δ, and dually. Collapsing all dead segments to points and letting G act in the obvious fashion on the resulting chain yields essentially the same ℓ-permutation group and exactly the same ℓ-group. We mention dead segments only in order to assume throughout that this collapsing has already been done, which causes (G,Ω) to have no dead segments. Groups having convex orbits automatically have no dead segments.

Recall that the transitive ℓ-permutation groups (G,Ω) which are *o-primitive* (i.e., lack proper convex congruences) are partitioned into three types:

(1) *Regular*, i.e., the regular representation of some subgroup of the additive reals $(\mathbb{R},+)$.

(2) *o-2-transitive*, i.e., for all $\alpha_1 < \beta_1$ and $\alpha_2 < \beta_2$ in Ω, there exists $g \in G$ such that $\alpha_1 g = \alpha_2$ and $\beta_1 g = \beta_2$.

(3) *Periodically o-primitive*, i.e., neither of the above. In this case, there exists $e < z \in A(\bar{\Omega})$ (the *period*) for which $\{\alpha z^n|\ n \in \mathbb{Z}\}$ is coterminal in $A(\bar{\Omega})$ for one (hence every) $\alpha \in \bar{\Omega}$, and for which the centralizer $\mathbb{Z}_{A(\bar{\Omega})}G$ is the cyclic subgroup generated by z. (Thus $(\omega z)g = (\omega g)z$.) Moreover, for $\alpha \in \bar{\Omega}$, and Δ the Ω-interval $(\alpha,\alpha z)$, $G_\alpha|\Delta$ acts o-2-transitively on Δ. (G,Ω) is *full* if $G = \mathbb{Z}_{A(\bar{\Omega})}z$. The canonical example is $G = \{g \in A(\mathbb{R}): (\omega + 1)g = \omega g + 1)$ for all $\omega \in \mathbb{R}\}$, with translation by 1 as the period z.

For o-2-transitive groups, we give here a property which says a bit more than the usual o-n-transitivity. For Π a segment of Ω, G_{Π} denotes $\{g \in G | \Pi g = \Pi\}$. Of course the action (G_{Π},Π) need not be faithful. Also, if (G,Ω) is periodically o-primitive with period z, we say that a segment Π of Ω is *at most one period long* if $\sup \Pi \leq (\inf \Pi)z$.

PROPOSITION 1.1. Let (G,Ω) be o-2-transitive, and let Π be an open segment of Ω. Then the action (G_{Π},Π) has the following multiple transitivity property: Given $n \in \mathbb{Z}^+$, and given any $\alpha_1 < \dots < \alpha_n \in \bar{\Pi}$ and $\beta_1 < \dots < \beta_n \in \bar{\Pi}$ with each $\beta_i \in \alpha_i G$, there exists $g \in G$ such that $\alpha_i g = \beta_i$ for all i.

Moreover, the same is true when (G,Ω) is transitive and periodically o primitive, provided Π is at most one period long.

Proof. For the o-2-transitive case, the proof is essentially the same as the proof of o-n-transitivity ([M, Lemma 4] or [G, Lemma 1.10.1]), $\inf \Pi$ and $\sup \Pi$ being treated as cuts in Ω which are to be fixed by g. The periodic case then follows from the fact that for $\sigma = \inf \Pi$, G_σ acts o-2-transitively on $(\sigma,\sigma z) \supseteq \Pi$. □

$\alpha,\beta \in \bar{\Omega}$ are said to be *tied* if $G_\alpha = G_\beta$, from which it follows that $\alpha g_1 = \alpha g_2$ iff $\beta g_1 = \beta g_2$ $(g_i \in G)$, so that where $g \in G$ sends α governs (and is governed by) where g sends β. If (G,Ω) is periodically o-primitive with period z, α and αz^n are tied (for any $\alpha \in \bar{\Omega}$, $n \in \mathbb{Z}$), and there are no other tyings. This is the archetypical example for our first definition of $G^{\cdot}$.

$G^{\cdot}$ is the set of all $h \in A(\bar{\Omega})$ which respect the tyings of (G,Ω) in the following sense. If $\alpha,\beta \in \bar{\Omega}$ are tied, then where h sends α governs (and is governed by) where h sends β according to the rule

$$\beta h = \sup\{\beta g | g \in G, \alpha g \leq \alpha h\}$$

$$= \inf\{\beta g | g \in G, \alpha g \geq \alpha h\}.$$

If $\alpha h \in \alpha G$, this condition simply says that βh is what it would have to be if h were in G, and thus α and β are tied in (G,Ω) iff they are tied in $(G^{\cdot},\bar{\Omega})$. However, the condition does not require that $\alpha h \in \alpha G$.

In the preceding example, the reader should convince himself in detail that $G^{:} = \mathbb{Z}_{A(\bar{\Omega})}z$ (the full group associated with G). In the regular case, $G^{:}$ is the regular representation of all of $\mathbb{R}$ (unless G is a discrete subgroup of $\mathbb{R}$, in which case $G^{:} = G$). In the o-2-transitive case, $G^{:} = A(\bar{\Omega})$ because distinct cuts $\alpha \neq \beta$ are never tied.

The definition of $G^{:}$ does not stipulate that $\alpha \neq \beta$, and the case in which $\alpha = \beta$ does have content—it forces $\alpha h \in \text{symcl}(\alpha G) = \{\omega \in \bar{\Omega} \mid \forall \omega < \sigma \in \bar{\Omega},\ \exists\, g \in G$ such that $\omega \leq \alpha g < \sigma$; and dually$\}$, even if no $\beta \neq \alpha$ is tied to α. (The reader worried about forgetting to treat this case will be cheered by the nonobvious fact that it doesn't really matter—changing the definition of $G^{:}$ by stipulating that $\alpha \neq \beta$ yields an equivalent definition.) Consequently, if α has an immediate predecessor or successor, $\alpha h \in \alpha G$.

Let α and β be tied. The well defined map $\alpha g \to \beta g$ $(g \in G)$, together with the "identity map" on G, provides an ℓ-permutation group isomorphism between the (not necessarily faithful) actions $(G,\alpha G)$ and $(G,\beta G)$. Thus corrresponding *holes* (proper Dedekind cuts) in αG and βG have the same stabilizers. For $h \in G^{:}$, if $\alpha h \notin \alpha G$, the hole in αG determined by αh is filled by a single element of $\bar{\Omega}$ (rather than by a nonsingleton o-block) because $\alpha h \in \text{symcl}(\alpha G)$, and similarly for βh, which is the corresponding hole in βG. Hence αh and βh are also tied. Anticipating the closure of $G^{:}$ under inverse (Proposition 1.4), it follows that αh and βh are tied iff α and β are tied.

LEMMA 1.2. Let Δ be a nonsingleton o-block of (G,Ω), and let $h \in G^{:}$. Then $\Delta h = \Delta g$ for some $g \in G$.

Proof. Let $\alpha = \inf \Delta < \sup \Delta = \beta$. Since $\alpha h = \inf\{\alpha g \mid g \in G,\ \alpha g \geq \alpha h\}$, there is some $g_0 \in G$ for which $\alpha h \leq \alpha g_0 < \beta h$. Since $\beta h = \sup\{\beta g \mid g \in G,\ \beta g \leq \beta h\}$ and since βG doesn't meet $(\alpha g_0, \beta g_0)$, we have $\beta g_0 \leq \beta h$. But since α and β are tied, the only way to have both $\alpha h \leq \alpha g_0$ and $\beta h \geq \beta g_0$ is to have equality in both cases, making $\Delta g_0 = \Delta h$. □

We write $(G^{\cdot})_\alpha$ as $G^{\cdot}_\alpha$.

PROPOSITION 1.3. If G_α fixes β, then $G^{\cdot}_\alpha$ also fixes β.

Proof. If $G_\alpha = G_\beta$, this is obvious from the definition of $G^{\cdot}$. If $G_\alpha \subset G_\beta$, the convexification $\Delta = \mathrm{Conv}(\alpha G_\beta)$ is an o-block of (G,Ω), and $G_\Delta = G_\beta$. Let $h \in G^{\cdot}_\alpha$. Then $\Delta h \in \Delta G$ by the lemma, and necessarily $\Delta h = \Delta$ since $\alpha \in \Delta h$. Letting $\sigma = \inf \Delta$, we have $\sigma h = \sigma$. Since σ and β are tied, $\beta h = \beta$. □

PROPOSITION 1.4. $G^{\cdot}$ is an ℓ-subgroup of $A(\bar{\Omega})$, $(G^{\cdot})^{\cdot} = G^{\cdot}$, and the ℓ-group $G^{\cdot}$ is laterally complete.

Proof. Let α and β be tied, and let $h, h_1 \in G^{\cdot}$. Since αh and βh are tied, we have

$$\beta(hh_1) = (\beta h)h_1 = \sup\{(\beta h)g|\ g \in G,\ (\alpha h)g \leq (\alpha h)h_1\}$$

$$= \sup\{(\sup\{\beta g_1|\ g_1 \in G,\ \alpha g_1 \leq \alpha h\})g|\ g \in G,\ (\alpha h)g \leq (\alpha h)h_1\}$$

$$= \sup\{\beta g_1 g|\ g_1, g \in G, ,\ \alpha g_1 \leq \alpha h,\ (\alpha h)g \leq (\alpha h)h_1\}$$

$$\leq \sup\{\beta g_2|\ g_2 \in G,\ \alpha g_2 \leq \alpha(hh_1)\}$$

$$\leq \inf\{\beta g_3|\ g_3 \in G,\ \alpha(hh_1) \leq \alpha g_3\} \leq \beta(hh_1).$$

This last inequality results from the dual of the argument just given, and tells us that all the items in this list are in fact equal, making $hh_1 \in G^{\cdot}$.

Next we show that $\beta h^{-1} \in \mathrm{symcl}(\beta G)$. If not, let Δ be the largest segment of $\bar{\Omega}$ containing βh^{-1} and missing βG. Δ is an o-block of (G,Ω) (we count $\{\omega\}$ as an o-block even when $\omega \in \bar{\Omega}\backslash\Omega$). Also, Δ is nonsingleton (except possibly in the case in which βh^{-1} has an immediate predecessor or successor, but then $\beta = (\beta h^{-1})h \in (\beta h^{-1})G$, and we are done). Now $\beta \in \Delta h = \Delta g$ for some $g \in G$ (by Lemma 1.2), so that $\beta g^{-1} \in \Delta$, a contradiction.

For $g \in G$, we can apply $gh \in G^{\cdot}$ to the tied points

$\alpha,\beta \in \Omega$ to obtain $\beta g \leq \beta h^{-1}$ iff $\beta gh \leq \beta$ iff $\alpha gh \leq \alpha$ iff $\alpha g \leq \alpha h^{-1}$. In light of the preceding paragraph, $\beta h^{-1} = \sup\{\beta g | \alpha g \leq \alpha h^{-1}\}$, and dually, placing h^{-1} in $G^{:}$.

Since $\alpha h \leq \alpha$ iff $\beta h < \beta$, $h \vee e \in G^{:}$, making $G^{:}$ an ℓ-subgroup of $A(\bar{\Omega})$. For similar reasons, the pointwise supremum of any pairwise disjoint set of elements of $G^{:}$ lies in $G^{:}$. It follows easily from the definition of $G^{:}$ that $(G^{:})^{:} = G^{:}$. □

Although we shall work here with the above definition of $G^{:}$, we shall use the following equivalent definitions to obtain alternate formulations of our results.

First, $G^{:}$ consists precisely of those $h \in A(\bar{\Omega})$ which can be *finitely gated*, meaning that given any $\alpha_1,\ldots,\alpha_n \in \bar{\Omega}$ and any gates $\langle\sigma_1,\tau_1\rangle,\ldots,\langle\sigma_n,\tau_n\rangle$ (with $\sigma_i,\tau_i \in \bar{\Omega}$ and $\sigma_i < \alpha_i h < \tau_i$), there exists $g \in G$ passing through all the gates, i.e., $\sigma_i < \alpha_i g < \tau_i$ for all i. The superscript in the $G^{:}$ notation is intended to be reminiscent of a gate, and $(G^{:},\bar{\Omega})$ will be called the *gate completion* of (G,Ω). Another way of saying this is that $G^{:}$ is the closure of G in the coarse stabilizer topology on $A(\bar{\Omega})$, which has as a generic basic open set the collection of all elements of $G^{:}$ passing through a given finite set of gates [B1].

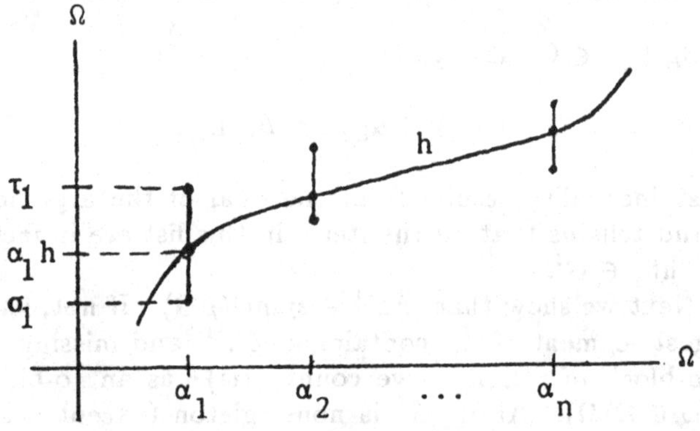

Figure 1

Equivalently (though not obviously so), one may build into this definition of $G^{:}$ the requirement that $\sigma_i,\tau_i \in \alpha_i A(\bar{\Omega}) = (\alpha_i h)A(\bar{\Omega})$. Because the closed primes of $A(\bar{\Omega})$ other than

$A(\bar{\Omega})$ itself are precisely the stabilizer subgroups of $(A(\bar{\Omega}), \bar{\Omega})$ [G, Theorem 8E], this means that $G^{:}$ is also the closure of G in the coarse closed prime topology on $A(\bar{\Omega})$ and in the α-topology on $A(\bar{\Omega})$, both of which are defined in ℓ-group language and are independent of the particular representation). These notions, which are discussed in [B1], will be used here solely to give alternate formulations of our results. For the record, and without proof:

THEOREM 1.5. All the preceding definitions of $G^{:}$ are equivalent.

2. Statement of results

Here we state three major theorems, only the third of which we shall prove. The proof, which does not depend on the first two theorems, will be given in §4.

THEOREM 2.1. All ℓ-group laws holding in G hold also in $G^{:}$. $G^{:}$ is order closed in $A(\bar{\Omega})$. $(G^{:},\bar{\Omega})$ has closed stabilizer subgroups, making $G^{:}$ completely distributive; and thus sups in $G^{:}$ are always pointwise, making $G^{:}$ a complete ℓ-subgroup of $A(\bar{\Omega})$. Moreover, G is order dense in $G^{:}$ if and only if (G,Ω) has closed stabilizer subgroups. Finally, if G is completely distributive and thus has a representation having closed stabilizer subgroups, $G^{:}$ is independent of the representation, i.e., all $G^{:}$'s resulting from closed-stabilizer representations are isomorphic over G, and in fact $G^{:} = G^{\alpha} = G^{i\alpha}$, the (iterated) α-completion of G.

Ordinarily there is no nice way to form an arbitrary element of the order closure of G in an ℓ-supergroup H (i.e., the smallest order closed ℓ-subgroup of H containing G). The next theorem is very unusual.

THEOREM 2.2. Let (G,Ω) be an ℓ-permutation group. Then

(1) Every $h \in G^{:}$ can be written as $h = \bigvee_i \bigwedge_j \bigvee_k g_{ijk}$ $(g_{ijk} \in G)$, and dually, with all indicated sups and infs lying

in $G^{:}$ and formed pointwise, thus being valid in both $A(\bar{\Omega})$ and $G^{:}$.

(2) The order closure of G in $G^{:}$ is all of $G^{:}$; and G is dense in the α-topology (= coarse stabilizer topology = coarse closed prime topology) on $G^{:}$.

(3) The order closure of G in $A(\bar{\Omega})$ is $G^{:}$, and the closure of G in the α-topology (= coarse stabilizer topology = coarse closed prime topology) on $A(\bar{\Omega})$ is $G^{:}$.

Part (1) is the most important result in this paper, although we shall forego proving it in order to prove Theorem 2.3. In (2) and (3), the claims about order closures follow from (1) given that $G^{:}$ is order closed in $A(\bar{\Omega})$, and the rest consists of reformulations utilizing the gating versions of $G^{:}$.

Can we improve on Theorem 2.2, writing each $h \in G^{:}$ as $h = \bigvee_i \bigwedge_j g_{ij}$ ($g_{ij} \in G$)? In general, no. There exists a completely distributive normal valued transitive ℓ-permutation group (G,Ω) for which there exists $h \in G^{:}$ which cannot be written as $h = \bigvee_i \bigwedge_j g_{ij}$ ($g_{ij} \in G$) or dually. However this *can* be done when Ω satisfies the *Souslin condition*, i.e., when every pairwise disjoint set of open intervals is countable—and every countable ℓ-group G can be represented on a countable Ω.

THEOREM 2.3. Let (G,Ω) be an ℓ-permutation group such that Ω satisfies the Souslin condition. Then every $h \in G^{:}$ can be written as $h = \bigvee_i \bigwedge_j g_{ij}$ ($g_{ij} \in G$), with all indicated sups and infs lying in $G^{:}$ and formed pointwise, thus being valid in both $A(\bar{\Omega})$ and $G^{:}$; and dually.

Because of the remarks in Theorem 2.1, Theorems 2.2 and 2.3 give information about α-completions G^{α} of completely distributive ℓ-groups G. In [B2, Corollary 1.6 ff.] the question is raised of whether every $h \in G^{\alpha}$ can be written as $h = \bigvee_i \bigwedge_j g_{ij}$ ($g_{ij} \in G$). Here we see that in the completely distributive case the answer is "almost" but "not quite", but that the answer is "yes" when G is countable.

3. Harmony

Although the rest of our results hold for arbitrary ℓ-permutation groups (G,Ω), and will mostly be stated that way, we shall assume in the proofs that G has convex orbits, i.e., that (G,Ω) is the splicing together of a collection of transitive actions of G. Every ℓ-group G can be represented in this way, using any Holland representation. If G is completely distributive, then there is a Holland representation which uses only closed primes and thus has closed stabilizer subgroups.

Recall that when (G,Ω) is transitive, the collection of convex congruences $\mathfrak{C}$ is totally ordered by refinement (finer being smaller). The tower of covering pairs of convex congruences is denoted by $\Gamma = \Gamma(G,\Omega)$. For each $\gamma \in \Gamma$, the transitive ℓ-permutation group $(G_{\Pi},\Pi/\mathfrak{C}_\gamma)$ is independent of the particular $\mathfrak{C}^\gamma$-class Π, and is o-primitive; it is denoted by (G_γ,Ω_γ), and is called an *o-primitive component* of (G,Ω). (G_γ,Ω_γ) must fit one of the three cases described in §1. For $\alpha \neq \beta \in \Omega$, there exists a unique $\gamma \in \Gamma$ (the *value* $\mathrm{Val}(\alpha,\beta)$) such that $\alpha\mathfrak{C}^\gamma\beta$ but not $\alpha\mathfrak{C}_\gamma\beta$.

When (G,Ω) has convex orbits, the actions (G_{Π},Π) for the various orbits Π of G are called the *constituents* of (G,Ω). $\Gamma(G,\Omega)$ means the union of the Γ's of the constituents, γ's from different constituents being incomparable. By the *o-primitive components* of (G,Ω) we mean those of its constituents. For $\alpha \neq \beta \in \Omega$ lying in the same constituent, $\mathrm{Val}(\alpha,\beta)$ refers to $\gamma = \mathrm{Val}(\alpha,\beta)$ in that constituent.

$G^{:}$ respects essentially all of the structure of (G,Ω). However, we saw in the o-primitive examples that $G^{:}$ can send points of Ω to holes, and thus fail to respect the trivial congruence (whose classes are singletons). Now that we want to view $G^{:}$ as having convex orbits, we enlarge Ω to $\Omega^{:} = \Omega G^{:} \subseteq \bar{\Omega}$, and consider $(G^{:},\Omega^{:})$ rather than $(G, \bar{\Omega})$. Of course, $\Omega^{:}$ is a union of G-orbits.

When (G_γ,Ω_γ) is regular, we say for short that γ is regular.

PROPOSITION 3.1. Suppose (G,Ω) has convex orbits. Then $G^{:}$ respects the nontrivial convex congruences of the constituents of (G,Ω), and thus $\Gamma(G^{\cdot},\Omega^{\cdot}) = \Gamma(G,\Omega)$. Moreover, each o-primitive component $(G^{\cdot}_{\gamma},\Omega^{\cdot}_{\gamma})$ has the same o-primitivity type as $(G_{\gamma},\Omega_{\gamma})$, and is obtained by enlarging $(G_{\gamma},\Omega_{\gamma})$ at most to:

(1) If γ is minimal, then

(a) If γ is regular, to $(\mathbb{R},\mathbb{R})$. (If γ is discrete, $(G^{\cdot}_{\gamma},\Omega^{\cdot}_{\gamma}) = (G_{\gamma},\Omega_{\gamma})$.)

(b) If γ is o-2-transitive, to $(A(\bar{\Omega}_{\gamma}),\Omega_{\gamma}A(\bar{\Omega}_{\gamma}))$.

(c) If γ is periodic with period z_{γ}, to $(\mathbb{Z}_{A(\bar{\Omega}_{\gamma})}z_{\gamma}, \Omega_{\gamma}\mathbb{Z}_{A(\bar{\Omega}_{\gamma})}z_{\gamma})$.

(2) If γ is nonminimal, then $\Omega^{\cdot}_{\gamma} = \Omega_{\gamma}$ and $G^{\cdot}_{\gamma}$ is at most the intersection with $A(\Omega_{\gamma})$ of the $G^{\cdot}_{\gamma}$ in (1).

Proof. The proposition follows from Lemma 1.2 and from the (reasoning underlying the) description of $G^{\cdot}$ in the o-primitive examples. The reason for the words "at most" is that even in the transitive case, a point ω can be tied to (the end points of) a nonsingleton o-block, forcing $\omega G^{\cdot} = \omega G$. □

Now let g be an o-automorphism of Ω and let $\omega \in \text{supp}(g) = \{\alpha g | \alpha \in \Omega, \alpha g \neq \alpha\}$. $\{\beta \in \Omega | \omega g^{-n} < \beta < \omega g^{n}$ for some $n \in \mathbb{Z}\}$ is called a *supporting interval* of g. By the *supporting intervals of* (G,Ω) (or for short, of G), we mean the supporting intervals of elements of G.

There are several possible relations of a supporting interval Δ to the structure of (G,Ω), depending on $\gamma = \text{Val}(\omega,\omega g)$, which is independent of $\omega \in \Delta$ and will be referred to as the value $\gamma(\Delta)$ of Δ:

(1) If γ is regular, $\Delta = \omega\mathfrak{C}^{\gamma}$.

(2) If γ is o-2-transitive, Δ is the Ω-interval corresponding to an interval (σ,τ) of $\omega\mathfrak{C}^{\gamma}/\mathfrak{C}_{\gamma}$, where σ and τ may be points, holes, or end points of $\omega\mathfrak{C}^{\gamma}/\mathfrak{C}_{\gamma}$.

(3) If γ is periodic with period z_{γ}, $\Delta = \omega\mathfrak{C}^{\gamma}$ unless Δ is at most one period long, in which case Δ fits the description in (2) with $\tau \leq \sigma z_{\gamma}$.

If for some $\omega \in \Omega$ and some o-2-transitive or periodic γ, (σ,τ) is an interval of $\omega\mathfrak{C}^{\gamma}/\mathfrak{C}_{\gamma}$ containing $\omega\mathfrak{C}_{\gamma}$, then Proposition 1.1 guarantees that $\omega\mathfrak{C}_{\gamma} \subseteq \Delta \subseteq (\sigma,\tau)$ for some supporting interval Δ of G.

A supporting interval of $G^{\cdot}$ need not be a supporting interval of G. For example, when $G = B(\mathbb{R})$, the o-automorphisms of bounded support, $G^{:} = A(\mathbb{R})$. However, any supporting interval of $(G^{:},\Omega^{:})$ must fit one of the possibilities for Δ listed above, either with reference to the structure of $(G^{:},\Omega^{:})$, or equivalently because of Proposition 3.1, with reference to the structure of (G,Ω) (provided Δ is enlarged to the corresponding $\Omega^{:}$-interval $\bar{\Delta} \cap \Omega^{:}$). Whenever we view a supporting interval Δ of G as a supporting interval of $G^{:}$, we should (but usually won't) speak of $\bar{\Delta} \cap \Omega^{:}$, thus perhaps filling in some holes of Δ. Thus $G^{:}_{\Delta}$ means $\{h \in G^{:} | (\bar{\Delta} \cap \Omega^{:})h = \bar{\Delta} \cap \Omega^{:}\} = \{h \in G^{:} | \bar{\Delta}h = \bar{\Delta}\}$. In this sense, every supporting interval of G is also a supporting interval of $G^{:}$. Similarly, when we view a supporting interval Δ of $G^{:}$ as an interval of Ω, we should speak of $\Delta \cap \Omega$; fortunately, $G_{\Delta} = G_{\Delta \cap \Omega}$.

For Δ_1 and Δ_2 supporting intervals of (G,Ω), or of $(G^{:},\Omega^{:})$, it may happen that for some $\alpha_i \in \bar{\Delta}_i$, α_1 and α_2 are tied (in G or equivalently in $G^{:}$). When this is so, we shall say that Δ_1 and Δ_2 are *harmonized* (in (G,Ω) or equivalently in $(G^{:},\Omega^{:})$).

The archetypical example is this: Let (G,Ω) be a periodically o-primitive group with period z, and let Δ be a supporting interval of some $g \in G$ which is at most one period long. Then Δ and Δz are harmonized.

For $\alpha \in \Delta$, we have $\alpha G_{\Delta} = \alpha G \cap \bar{\Delta}$, for if $\alpha \leq \alpha g \in \bar{\Delta}$ there exists $g_1 \in G$ such that $(\alpha g)g_1 = \alpha g$ and $\Delta g g_1 = \Delta$ (using Proposition 1.1 in some cases), and dually. Similarly, $\alpha G^{:}_{\Delta} = \alpha G^{:} \cap \bar{\Delta}$.

When Δ_1 and Δ_2 are harmonized supporting intervals of $G^{:}$, with $\alpha_1 \in \bar{\Delta}_1$ and $\alpha_2 \in \bar{\Delta}_2$ tied, an ℓ-permutation group isomorphism between the actions $(G,\alpha_1 G)$ and $(G,\alpha_2 G)$ is given by $\alpha_1 g \leftrightarrow \alpha_2 g$ $(g \in G)$ and the identity map on G, making $G_{\Delta_1} = G_{\Delta_2}$. Then in light of the previous paragraph, an ℓ-permutation group isomorphism between $(G_{\Delta_1},\alpha_1 G_{\Delta_1})$ and $(G_{\Delta_2},\alpha_2 G_{\Delta_2})$ is given by $\alpha_1 g \leftrightarrow \alpha_2 g$ $(g \in G_{\Delta_1} = G_{\Delta_2})$ and the identity map on $G_{\Delta_1} = G_{\Delta_2}$. Similar comments apply to $G^{:}$. This is the reason for the term "harmonized". But beware—since a point can be *tied* to a nonsingleton o-block, and $\gamma(\Delta_1)$ and $\gamma(\Delta_2)$ need not coincide. (As with elements of $\bar{\Omega}$, we say that two o-blocks are *tied* if their stabilizer subgroups coincide.)

In the preceding paragraph, either α_i is contained in (the completion, including end points, of) some $\mathcal{C}_{\gamma_i}$-class $\alpha_i\mathcal{C}_{\gamma_i}$ (where $\gamma_i = \gamma(\Delta_i)$), or α_i is a hole in $\alpha_i\mathcal{C}^{\gamma_i}/\mathcal{C}_{\gamma_i}$ which we denote by $\alpha_i\mathcal{C}_{\gamma_i}$, and Δ_i is related to $\alpha_i\mathcal{C}^{\gamma_i}$ as described above. $\alpha_1\mathcal{C}_{\gamma_1}$ and $\alpha_2\mathcal{C}_{\gamma_2}$ must be tied because their stabilizer subgroups are both values of g containing $G_{\alpha_1} = G_{\alpha_2}$. Letting $\Pi_i = \alpha_i\mathcal{C}^{\gamma_i}$, an isomorphism between the actions $(G_{\Pi_1}, \overline{\Pi_1/\mathcal{C}_{\gamma_1}})$ and $(G_{\Pi_2}, \overline{\Pi_2/\mathcal{C}_{\gamma_2}})$ is given by $\alpha_1\mathcal{C}_{\gamma_1}g \leftrightarrow \alpha_2\mathcal{C}_{\gamma_2}g$ $(g \in G_{\Pi_1} = G_{\Pi_2})$ and the identity map on $G_{\Pi_1} = G_{\Pi_2}$. In particular, γ_1 and γ_2 must have the same o-primitivity type.

PROPOSITION 3.2. Let Δ_1 and Δ_2 be supporting intervals of G (or, more generally, of $G^{\dot{:}}$). If Δ_1 and Δ_2 are harmonized, then for any $\alpha_1 \in \bar{\Delta}_1$, there exists $\alpha_2 \in \bar{\Delta}_2$ such that $G_{\alpha_1} \subseteq G_{\alpha_2}$ (equivalently, $G^{\dot{:}}_{\alpha_1} \subseteq G^{\dot{:}}_{\alpha_2}$), and α_2 may be taken to be a point or hole in $\Delta_2/\mathcal{C}_{\gamma_2}$, where $\gamma_2 = \gamma(\Delta_2)$. Conversely, if there exist $\alpha_i \in \bar{\Delta}_i$ such that $G_{\alpha_1} \subseteq G_{\alpha_2}$, then Δ_1 and Δ_2 are harmonized provided they are both supporting intervals of some one $h \in G^{\dot{:}}$.

Proof. Suppose Δ_1 and Δ_2 are harmonized, and let $\alpha_1 \in \bar{\Delta}_1$. If possible, pick $\alpha_2 \in \bar{\Delta}_2$ tied to α_1. But suppose there is no such α_2. Then pick $\beta_1 \in \bar{\Delta}_1$ tied to some $\beta_2 \in \bar{\Delta}_2$. $\alpha_1 \notin \text{symcl}(\beta_1 G)$, else α_1 would be tied to the corresponding cut in $\beta_2 G$. Let Λ_1 be the largest segment of Ω which contains α_1 and misses $\beta_1 G$. Then Λ_1 is an o-block of (G,Ω) filling in a hole in the chain $\beta_1 G \cap \bar{\Delta}_1$. Let σ_2 be the corresponding hole in $\beta_2 G \cap \bar{\Delta}_2$. (If that hole is filled in with an o-block Λ_2, let σ_2 be either end point of Λ_2.) Then $G_{\alpha_1} \subseteq G_{\Lambda_1} = G_{\sigma_2} \subseteq G_{\alpha_2}$ for some point or hole α_2 in $\Delta_2/\mathcal{C}_{\gamma_2}$. By Proposition 1.3, $G_{\alpha_1} \subseteq G_{\alpha_2}$ iff $G^{\dot{:}}_{\alpha_1} \subseteq G^{\dot{:}}_{\alpha_2}$.

Conversely, if $G_{\alpha_1} \subseteq G_{\alpha_2}$, then $\text{Conv}(\alpha_1 G_{\alpha_2})$ is an o-block of (G,Ω). Letting β_1 be either of its end points, we have $G_{\beta_1} = G_{\alpha_2}$ and thus also $G^{\dot{:}}_{\beta_1} = G^{\dot{:}}_{\alpha_2}$. $\beta_1 \in \bar{\Delta}_1$ provided Δ_1 and Δ_2 are both supporting intervals of some one $h \in G^{\dot{:}}$, for $\alpha_2 \in \bar{\Delta}_2$ and thus $\beta_1 \in \bar{\Delta}$ for some supporting interval Δ of h, and necessarily $\Delta = \Delta_1$. □

COROLLARY 3.3. The harmonic relation ($\Delta_1 \sim \Delta_2$ iff Δ_1 and Δ_2 are harmonized) is an equivalence relation on the set of supporting intervals of $G^{:}$.

Proof. Suppose Δ_1 and Δ_2 are harmonized, and also Δ_2 and Δ_3. Then $G_{\Delta_1} = G_{\Delta_2} = G_{\Delta_3}$. Now pick $\beta_2 \in \bar{\Delta}_2$ and $\beta_3 \in \bar{\Delta}_3$ which are tied. By Proposition 3.2, there exists $\beta_1 \in \bar{\Delta}_1$ such that $G_{\beta_2} \subseteq G_{\beta_1}$. Let η_i be either end point of $\mathrm{Conv}(\beta_i G_{\beta_1})$, $i = 2,3$. Then $G_{\eta_3} = G_{\eta_2} = G_{\beta_1}$. Since Δ_1 and Δ_2 are harmonized and Δ_2 meets $\mathrm{Conv}(\beta_2 G_{\beta_1})$, $\eta_2 \in \bar{\Delta}_2$. Since Δ_2 and Δ_3 are harmonized and Δ_3 meets $\mathrm{Conv}(\beta_3 G_{\eta_2})$, $\eta_3 \in \bar{\Delta}_3$. □

For Δ a supporting interval of $G^{:}$, we denote its equivalence class under the harmonic relation by $\mathrm{Harm}(\Delta)$. Each $\Delta' \in \mathrm{Harm}(\Delta)$ will be called a *harmonic* of Δ. If Δ is a supporting interval of a particular $h \in G^{:}$, every harmonic of Δ is also a supporting interval of that h.

(G,Ω) is called *depressible* if for any $g \in G$ and any supporting interval Δ of g, the result h of depressing g off Δ (meaning that $\omega h = \omega g$ for $\omega \in \Delta$ but $\omega h = \omega$ for $\omega \in \Omega\backslash\Delta$) lies in G.

We shall call (G,Ω) *harmonically depressible* if for any $g \in G$ and any supporting interval Δ of g, the result g^{Δ} of depressing g off $\bigcup \mathrm{Harm}(\Delta)$ lies in G. Depressible ℓ-permutation groups are automatically harmonically depressible since each $\mathrm{Harm}(\Delta)$ is singleton.

For o-2-transitive and regular o-primitive ℓ-permutation groups, these two notions coincide (and automatically obtain in the second case). In contrast, periodically o-primitive ℓ-permutation groups are never depressible, but may well be harmonically depressible. In fact, full periodically derived ℓ-permutation groups are automatically harmonically depressible.

Indeed, almost all o-primitive groups which occur in nature (except pathologically o-2-transitive groups) are harmonically depressible, and it is easily checked that forming generalized wreath products preserves harmonic depressibility.

THEOREM 3.4. $(G^{:},\Omega^{:})$ is harmonically depressible.

Proof. For Δ a supporting interval of $h \in G^{\cdot}$, h^{Δ} satisfies the defining condition for membership in $(G^{\cdot})^{\cdot} = G^{\cdot}$ since for tied $\alpha,\beta \in \bar{\Omega}$, either both or neither lie in $\bigcup \mathrm{Harm}(\Delta)$. □

4. Proof of Theorem 2.3

Given pairwise disjoint supporting intervals of $g_1,\dots,g_n$, to what extent can the actions of the g_i's on finitely many designated cuts in their respective supporting intervals be matched by some one $g \in G$?

LEMMA 4.1. Let (G,Ω) be an ℓ-permutation group. Let $g_1,\dots,g_n$ be not necessarily distinct elements of G, and let $\Delta_1,\dots,\Delta_n$ be pairwise disjoint supporting intervals of $g_1,\dots,g_n$. Let $\alpha_1,\dots,\alpha_m \in \bar{\Omega}$.

(1) Suppose that no G_{σ_j} $(\sigma_j \in \bar{\Delta}_j)$ fixes any $\sigma_k \in \bar{\Delta}_k$ for which $g_k \neq g_j$, and no G_{α_j} fixes any $\sigma_k \in \bar{\Delta}_k$. Then given any finite sets $\delta_{i,1},\dots,\delta_{i,p_i} \subseteq \bar{\Delta}_i$ $(i = 1,\dots,n)$, there exists some one $g \in G$ which for each i agrees with g_i at $\delta_{i,1},\dots,\delta_{i,p_i}$, and which fixes $\alpha_1,\dots,\alpha_m$.

(2) Statement (1) also holds when the σ_i's and $\delta_{i,n}$'s are taken to be points and holes in $\Delta_i/\mathfrak{C}_{\gamma_i}$ $(\gamma_i = \gamma(\Delta_i))$.

Proof. We prove (1), the proof of (2) being virtually identical. We assume temporarily that $g_1,\dots,g_n$ are distinct.

Let $\Pi_i = \{\delta_{i,1},\dots,\delta_{i,p_i}\} \cup \{\delta_{i,1},\dots,\delta_{i,p_i}\}g_i \subseteq \bar{\Delta}_i$. For $\sigma \in \Pi_j$ (with $j \neq i$) or $\sigma = \alpha_j$, G_σ fixes no cut in Δ_i (which extends beyond Π_i in both directions). Hence there exists $k_\sigma \in G_\sigma^+$ such that $(\inf \Pi_i)k_\sigma > \sup \Pi_i$, else $\sup\{(\inf \Pi_i)k_\sigma \mid k_\sigma \in G_\sigma^+\}$ would be a cut in Δ_i fixed by G_σ. Let f_i be the inf of these finitely many k_σ's, so that f_i fixes all these σ's and exceeds g_i throughout Π_i. Let $d_i = (g_i \wedge f_i) \vee f_i^{-1} \in G$, which agrees with g_i on Π_i and fixes all these σ's. Now take $g = d_1 d_2 \dots d_n$.

It should be clear how to modify the preceding argument to cope with nondistinct $g_1,\dots,g_n$ by amalgamating the various Π_i's whose g_i's coincide to produce a single d. □

LEMMA 4.2. Let Δ be a supporting interval of (G,Ω), $\gamma = \gamma(\Delta)$, and $g_1,g_2 \in G$. If $\xi g_1 < \xi g_2$ for every point (i.e., $\mathcal{C}_\gamma$-class) and hole ξ in $\Delta/\mathcal{C}_\gamma$, then $g_1 < g_2$ throughout $\bigcup \mathrm{Harm}(\Delta)$.

Proof. Let $\alpha_1 \in \bar{\Delta}_1$, $\Delta_1 \in \mathrm{Harm}(\Delta)$. By Proposition 3.2, $G_{\alpha_1} \subseteq G_\xi$ for some point or hole ξ in $\Delta/\mathcal{C}_\gamma$. Let $\Lambda = \mathrm{Conv}(\alpha_1 G_\xi)$, an o-block of (G,Ω) for which $G_\Lambda = G_\xi$. Since $\xi g_1 < \xi g_2$, $\Lambda g_1 < \Lambda g_2$ and thus $\alpha_1 g_1 < \alpha_1 g_2$. (Knowing merely that $\xi g_1 \leq \xi g_2$ would not suffice here.) □

Let $\{g_i | \ i \in I\}$ be o-automorphisms of a chain Ω. The *pointwise supremum* $f: \Omega \to \bar{\Omega} \cup \{+\infty\}$ is defined by $\alpha f = \sup_i \alpha g_i$. Certainly f is nondecreasing. If f is one-to-one and the range is a dense subset of $\bar{\Omega}$, so that f has a unique extension to an element of $A(\bar{\Omega})$, this extension will also be denoted by f; and conveniently $\alpha f = \sup_i \alpha g_i$ even for $\alpha \in \bar{\Omega}\backslash\Omega$. Under these conditions, $f = \bigvee g_i$ in $A(\bar{\Omega})$, the sup obviously being pointwise in the usual sense, and we write $f = \overset{p}{\bigvee} g_i$. The following lemma will be a crucial tool, providing conditions under which $\overset{p}{\bigvee} g_i$ exists. Its conditions (1) and (2) should be visualized in terms of the graphs of the g_i's.

LEMMA 4.3. Let $\{g_i | \ i \in I\}$ be o-automorphisms of a chain Ω. Then $\overset{p}{\bigvee} g_i$ exists and is an o-automorphism of Ω (not just of $\bar{\Omega}$), and for any (G,Ω) for which $\{g_i | \ i \in I\} \subseteq G$ we have $\overset{p}{\bigvee} g_i \in G^{:}$, all provided that

(1) For each $\alpha \in \bar{\Omega}$ there exists a finite subset $I_\alpha \subseteq I$ such that only the g_i's with $i \in I_\alpha$ are involved at α, i.e., $\alpha g_k \leq \max_{i \in I_\alpha} \alpha g_i$ whenever $k \in I\backslash I_\alpha$; and

(2) For each $\beta \in \bar{\Omega}$ there exists a finite subset $J_\beta \subseteq I$ such that only the g_i^{-1}'s with $i \in J_\beta$ are involved at β, i.e., $\beta g_k^{-1} \geq \min_{i \in J_\beta} \beta g_i^{-1}$ whenever $k \in I\backslash J_\beta$.

Proof. Let $f = \overset{p}{\bigvee} g_i$. By condition (1), $f: \Omega \to \Omega$. For $\alpha_1 < \alpha_2 \in \Omega$ only $\{g_i | \ i \in I_{\alpha_1} \cup I_{\alpha_2}\}$ are involved, so $\alpha_1 f < \alpha_2 f$. Now we show that f maps Ω *onto* Ω. Given $\beta \in \Omega$, (2) guarantees that for some $i \in J_\beta$, $\beta g_i^{-1} \leq \beta g_k^{-1}$ for *every* $k \in I$. Then $(\beta g_i^{-1})f = (\beta g_i^{-1})g_i = \beta$, for if $(\beta g_i^{-1})g_j > (\beta g_i^{-1})g_i$

for some $j \in I$, then $\beta g_j^{-1} < \beta g_i^{-1}$. Therefore f is an o-automorphism of Ω.

To show that $f \in G^{\cdot}$, consider points $\mu,\nu \in \bar{\Omega}$ for which $G_\mu = G_\nu$. By condition (1), there is some $i \in I$ such that $\mu f = \mu g_i$. Since $\mu g_i \geq \mu g_j$ for all $j \in I$ and since μ and v are tied, we have $\nu g_i \geq \nu g_j$ for all $j \in I$, making $\nu f = \nu g_i$. In fact, the same argument shows that for *any* $g \in G$, μg and μf are related exactly as are νg and νf. □

Proof of Theorem 2.3. For convenience, we prove the dual. Throughout, we make implicit use of Proposition 3.1.

Let $h \in G^{\cdot}$. Let $\alpha \in \Omega$ and $\alpha h \leq \beta \in \alpha G$. It will suffice to produce a pointwise sup $f = f_{\alpha\beta} = \overset{p}{\bigvee} g_j = \overset{p}{\bigvee} g_{\alpha\beta j} \in G^{\cdot}$ $(g_j = g_{\alpha\beta j} \in G)$ such that $\alpha f_{\alpha\beta} = \beta$ and $f_{\alpha\beta} \geq h$. Then $h = \underset{p}{\bigwedge} f_{\alpha\beta}$.

We may assume that:

(i) $\underline{\alpha h = \alpha}$. (For then in general, pick $g \in G$ such that $\alpha g = \beta$. Replacing h by $h \vee g$, we may assume that $\alpha h = \beta$. Then $\alpha h g^{-1} = \alpha$, and having found g_j's such that $\overset{p}{\bigvee} g_j \in G^{\cdot}$ and $\overset{p}{\bigvee} g_j$ fixes α and exceeds hg^{-1}, we take $f_{\alpha\beta} = \overset{p}{\bigvee} g_j g$, this last sup also being pointwise.) Now since $\alpha h = \alpha \in \alpha G$, we need merely treat the case $\beta = \alpha$, i.e., make $\alpha f = \alpha$.

(ii) $\underline{e < h \in G^{\cdot}_{\alpha}}$. (Replace h by h^+.)

(iii) $\underline{\text{No } e < h' \in h^{\perp} \text{ fixes } \alpha}$. (Among the pairwise disjoint subsets of $G^{\cdot}$ that contain h and fix α, pick a maximal one $\mathcal{H}$, and replace h by the pointwise sup of $\mathcal{H}$, which lies in $G^{\cdot}$ by the lateral completeness of $G^{\cdot}$.)

Now let $\Lambda = \text{supp}(h) \cup \{\alpha\}$, and let $\hat{\Lambda}$ consist of those $\omega \in \Omega$ such that $G_\lambda \subseteq G_\omega$ (i.e., G_λ fixes ω) for some $\lambda \in \Lambda$. If $G_{\omega_1} = G_{\omega_2}$, then $\omega_1 \in \hat{\Lambda}$ iff $\omega_2 \in \hat{\Lambda}$. We claim that $\hat{\Lambda}$ is dense in $\bar{\Omega}$.

First, $\hat{\Lambda}$ must meet the completion $\bar{\Delta}$ of every supporting interval Δ of every $g \in G$. Otherwise $\hat{\Lambda}$ would be disjoint from the completion $\bar{\Delta}'$ of every harmonic Δ' of Δ by Proposition 3.2, and depressing $g^{\pm 1}$ off $\text{Harm}(\Delta)$ would yield an $h' \in (G^{\cdot})^+_\alpha$ which violates (iii). Consequently $\hat{\Lambda}$ meets every nonsingleton o-block of (G,Ω).

Now let Π be a nonsingleton segment of $\bar{\Omega}$, and suppose first that Π contains more than two elements. Then we may pick an interior point π of Π lying in Ω. If $\{\pi\}$ is the intersection of the tower of nonsingleton o-blocks Δ containing it, then for some such Δ we have $\Pi \supseteq \Delta$, and then $\hat{\Lambda}$ meets Δ and thus Π. If not, then taking advantage of our assumption that (G,Ω) has convex orbits, the transitive action $(G_{\pi G}|\pi G, \pi G)$ has a smallest proper convex congruence $\mathcal{C}$, and with no loss of generality $\Pi \subseteq \pi\mathcal{C}$. Let $\Delta = \pi\mathcal{C}$. If $G_\Delta|\Delta$ is o-2-transitive or periodic, then Π contains some supporting interval of G (see the discussion of supporting intervals following the definition), so $\hat{\Lambda}$ meets Π. If $G_\Delta|\Delta$ is regular, then Δ itself is a supporting interval of G, so $\hat{\Lambda}$ meets Δ; and in fact $\hat{\Lambda}$ meets Π because all points in Δ are tied to one another. Finally, if Π contains just two points, the argument is like that in the regular case, $G_\Delta|\Delta$ being $\mathbb{Z}$. This completes the proof that $\hat{\Lambda}$ is dense in $\bar{\Omega}$.

Our goal is to produce $f = \overset{p}{\vee} g_j \in G^{\cdot}$ with each $g_j \in G_\alpha$ (making $\alpha f = \alpha$), and so that for each $\eta \in \Omega$, $\eta g_j \geq \eta h$ for some j (making $f \geq h$). We shall do this by enumerating a carefully chosen countable set $g_0, g_1, g_2, \ldots$ and applying Lemma 4.3 to guarantee that this set has a pointwise sup f. There will be a countable set $\mathcal{S} = \{s_0, s_1, s_2, \ldots\}$ of "places" at which we shall need to make the action of f sufficiently large and yet not *too* large. Roughly speaking, we shall make $s_j g_j$ lie in the appropriate range, make $s_j g_k$ smaller when $k > j$, and not worry about $s_j g_k$ for the finitely many $k < j$. Lemma 4.1 will guarantee the consistency of the specifications about any one g_j.

We form $\mathcal{S}$ as follows:

(1) $\alpha \in \mathcal{S}$.

(2) From each harmonic class of supporting intervals of h, pick one supporting interval $\Delta \subseteq \Omega^{\cdot}$ and an arbitrary $\delta \in \Delta \cap \Omega$. Precisely one of the following cases will obtain:

(a) If for some regular γ, Δ is a $\mathcal{C}^\gamma$-class, put Δ itself in $\mathcal{S}$.

(b) If for some o-2-transitive γ there is a $\mathcal{C}^\gamma$-class Σ with $\Delta/\mathcal{C}_\gamma$ a nonsingleton segment of $\Sigma/\mathcal{C}_\gamma$, let $\beta_0 = \beta_0^\Delta$ denote $\delta\mathcal{C}_\gamma \in \Delta/\mathcal{C}_\gamma$, let $\beta_i = \beta_0 h^i$, and put each pair $<\beta_i, \beta_{i+1}>$ in $\mathcal{S}$. The β_i's are points in $\Delta/\mathcal{C}_\gamma$ (not in Ω), and are coterminal in $\Delta/\mathcal{C}_\gamma$; see Figure 2. Because the β_i's need not all lie in the same G-orbit (in the case of minimal

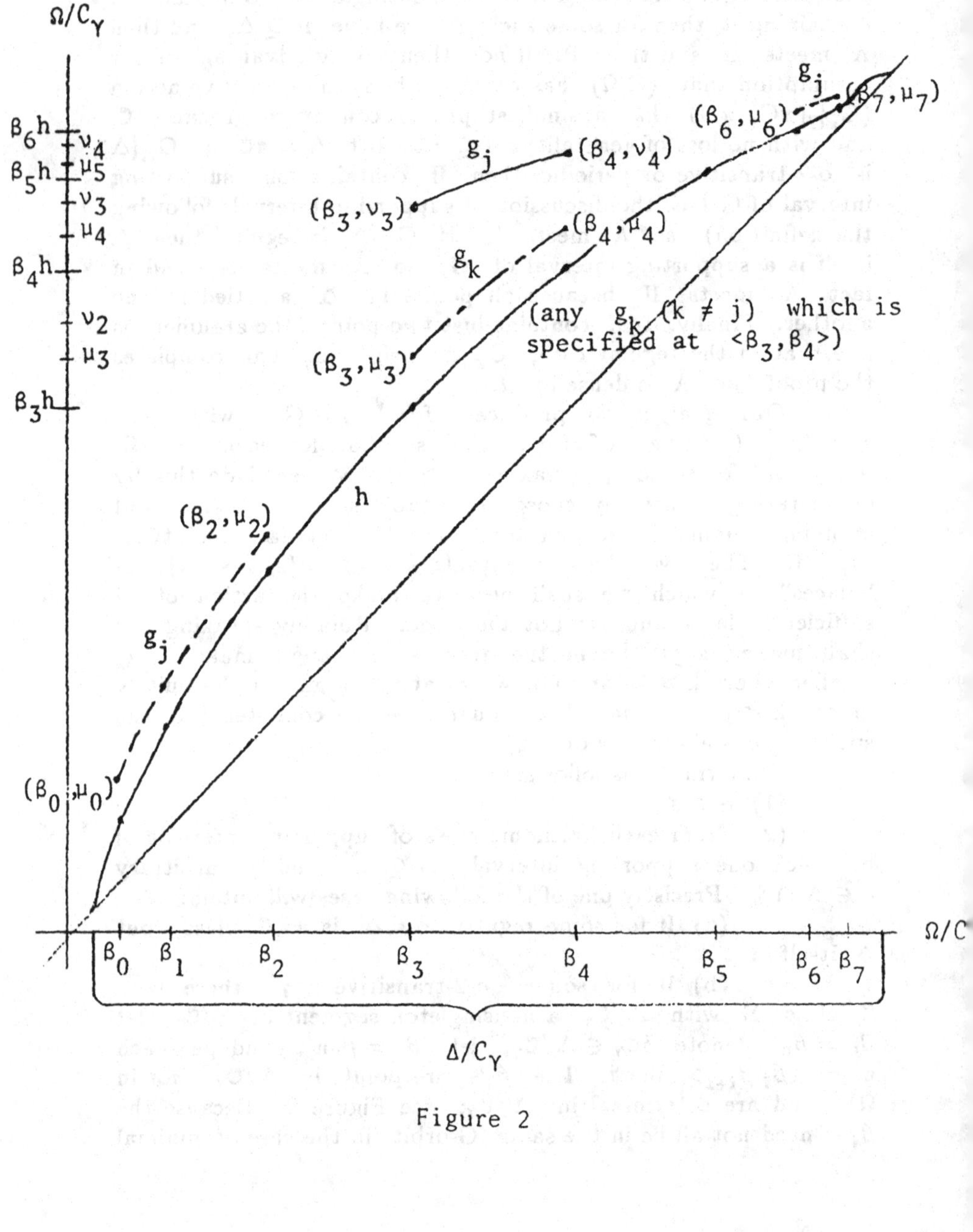

Ω/C_γ
$\beta_6 h$
ν_4
$\beta_5 h$
μ_5
ν_3
μ_4
$\beta_4 h$
ν_2
μ_3
$\beta_3 h$
g_j
(β_4,ν_4)
(β_3,ν_3)
(β_4,μ_4)
g_k
(any g_k $(k \neq j)$ which is
specified at <β_3,β_4>)
(β_3,μ_3)
h
(β_2,μ_2)
g_j
(β_0,μ_0)
g_j
(β_6,μ_6)
(β_7,μ_7)
β_0
β_1
β_2
β_3
β_4
β_5
β_6
β_7
Ω/C
Δ/C_γ

Figure 2

γ), we pick two subchains of $\Delta/\mathcal{C}_\gamma$ interspersed among the β_i's, choosing $\mu_i,\nu_i \in \Delta/\mathcal{C}_\gamma \cap \beta_i G$ such that $\beta_i h < \mu_i < \nu_{i-1} < \beta_{i+1}h$.

(c) If for some periodic γ there is a $\mathcal{C}^\gamma$-class Σ with $\Delta/\mathcal{C}_\gamma$ a nonsingleton segment of $\Sigma/\mathcal{C}_\gamma$, then

(i) If $\Delta = \Sigma$, put Δ in $\mathcal{S}$.

(ii) If not, so that $\Delta/\mathcal{C}_\gamma$ is at most one period long, proceed as in (2b), putting each $<\beta_i,\beta_{i+1}>$ in $\mathcal{S}$ and picking μ_i's and ν_i's as before.

By hypothesis Ω satisfies Souslin's condition, making $\mathcal{S}$ countable. We enumerate it as $s_0,s_1,s_2,\ldots$, with $s_0 = \alpha$.

Now we specify the action at s_j of $g_j,g_{j+1},g_{j+2},\ldots$, but not of $g_0,\ldots,g_{j-1}$.

(1′) For all j, αg_j $(= s_0 g_j) = \alpha$. (This forces $\eta g_j = \eta$ for all $\eta \in \Omega$ tied to α.)

(2′) In (2) above, let h_γ denote the action of h on $\Delta/\mathcal{C}_\gamma$.

(a′) When $s_j = \Delta$ fits case (2a), h_γ is in effect (translation by) a positive real number. We pick δ_1 so that $(\delta\mathcal{C}_\gamma)h < \delta_1\mathcal{C}_\gamma \in (\delta\mathcal{C}_\gamma)G \cap \Delta/\mathcal{C}_\gamma$, and specify that

$$(\delta\mathcal{C}_\gamma)g_j = \delta_1\mathcal{C}_\gamma.$$

We specify also that

$$(\delta\mathcal{C}_\gamma)g_k = \delta\mathcal{C}_\gamma \text{ when } k > j.$$

(These specifications force $\eta g_j > \eta h$, and $\eta g_k < \eta g_j$ when $k > j$, for all $\eta \in \bigcup\mathrm{Harm}(\Delta)$. This is because $\beta g_k < \beta g_j > \beta h$ whenever β is any point or hole in $\Delta/\mathcal{C}_\gamma$.)

(b′) When $s_j = <\beta_i,\beta_{i+1}>$ arises from some Δ fitting case (2b), we specify that

$$s_j g_j = <\nu_i,\nu_{i+1}>$$

(meaning that $\beta_i g_j = \nu_i$ and $\beta_{i+1}g_j = \nu_{i+1}$). We specify that

$$s_j g_k = <\mu_i,\mu_{i+1}> \text{ when } k > j,$$

except if $s_k = <\beta_{i-1},\beta_i>$ or $<\beta_{i+1},\beta_{i+2}>$ or $<\beta_{i+2},\beta_{i+3}>$, we omit this specification. This forces the graph of g_j to be "high" between β_i and β_{i+1}, and the graphs of the g_k's ($k > j$) to be "low" there except in the three cases where no specification is made there for g_k. See Figure 2, which shows the action on $\Delta/\mathcal{C}_\gamma$ of the g_j associated with $<\beta_3,\beta_4>$, i.e., $s_j = <\beta_3,\beta_4>$. Of course, g_j is specified only at certain β_i's, not at points in between. The portions of the graph indicated by dashes of uniform length indicate that g_j is defined at a particular $<\beta_i,\beta_{i+1}>$ only if $<\beta_i,\beta_{i+1}>$ precedes $<\beta_3,\beta_4>$ in the enumeration of $\mathcal{S}$.

The specifications in (2b′) force $\eta g_j > \eta h$, and also force $\eta g_k < \eta g_j$ when $k > j$ and we do not have one of the three cases causing omission of the specification for g_k, for all $\eta \in \bigcup \mathrm{Harm}(\Psi)$, where Ψ is the Ω-interval corresponding to the $(\Delta/\mathcal{C}_\gamma)$-interval $[\beta_i,\beta_{i+1}]$. This is because $\beta g_k < \beta g_j > \beta h$ whenever β is any point or hole in $\Psi/\mathcal{C}_\gamma$, as is evident from Figure 2. (There is a technical difficulty here, because if Ψ is not a supporting interval of $G^{\cdot}$, $\mathrm{Harm}(\Psi)$ is undefined. Fortunately, because of o-n-transitivity $\Psi = \bigcup\{\Psi' | \Psi' \subseteq \Psi$ and Ψ' is a supporting interval of $G^{\dot{}}\}$, and we construe $\mathrm{Harm}(\Psi)$ to mean $\bigcup\{\mathrm{Harm}(\Psi') | \Psi' \subseteq \Psi$ and Ψ' is a supporting interval of $G^{\dot{}}\}$.)

(c′) When Δ fits (2c):

(i′) In case (i), we proceed essentially as in (2a′), arranging that $(\delta\mathcal{C}_\gamma)g_j > (\delta\mathcal{C}_\gamma)hz_\gamma$ to make sure that $\xi g_j > \xi h$, and that $\xi g_k < \xi g_j$ when $k > j$, whenever ξ is any point or hole in $\Delta/\mathcal{C}_\gamma$.

(ii′) In case (ii), we proceed as in (2b′).

Does there really exist $g_j \in G$ meeting the above specifications? The specifications for g_j were made only at the finitely many elements $s_0,\ldots,s_j$ (with up to three of these omitted). For any one supporting interval Δ of h, all the specifications made within Δ for g_j can be simultaneously achieved by some one g_j^Δ (because of Proposition 1.1, and because of the omitted specifications in (2b′) and (2c,i′)—cf. Figure 2).

In cases (2a′) and (2c,i′), Δ itself is a supporting interval of g_j^Δ. In cases (2b′) and (2c,ii′), we want to arrange that this be nearly true. Let β_m be the smallest β_i in Δ at which g_j was specified, and β_M the largest. For any β_i ($i = m,...,M$) for which g_j was not specified, we can arrange that $\beta_i g_j^\Delta \geq \mu_i > \beta_{i+1}$ (so that $\beta_m,...,\beta_M$ all lie within one supporting interval Δ_1 of g_j^Δ), and that $\beta_{m-1}g_j^\Delta = \beta_{m-1}$ and $\beta_{M+3}g_j^\Delta = \beta_{M+3}$ (so that $\Delta_1 \subseteq \Delta$). The hypotheses of part (2) of Lemma 4.1 are satisfied by the finitely many g_j^Δ's because distinct harmonized Δ's are never both included in $\mathcal{S}$, and because every $\beta \in \bar{\Delta}_1 \subseteq \bar{\Delta}$ is moved by some element of $G_\alpha^{:}$ (namely the given h) and thus by some element of G_α because of Proposition 1.3. Now Lemma 4.1 guarantees the existence of the desired g_j. Replacing g_j by g_j^+, we may assume that $g_j \in G_\alpha^+$.

We want to apply Lemma 4.3 to $g_0,g_1,g_2,...$ (the index set I being the natural numbers), so we verify its hypotheses.

We set $I_\alpha = \{0\} = J_\alpha$, and also $I_\omega = \{0\} = J_\omega$ for any ω fixed by G_α.

Next, let $\eta \in \text{supp}(h)$. Then $\eta \in \Delta'$ for some harmonic Δ' of some Δ included in $\mathcal{S}$. If Δ satisfies (2a) or (2c,i), so that Δ itself is an element s_j of $\mathcal{S}$, set $I_\eta = J_\eta = \{0,1,...,j\}$; cf. the remark about "$\eta g_k < \eta g_j$ when $k > j$" in (2a′). If Δ satisfies (2b) or (2c,ii), then η lies in a harmonic of the Ω-interval corresponding to the $(\Delta/\mathcal{C}_\gamma)$-interval $[\beta_i,\beta_{i+1}]$ for some $<\beta_i,\beta_{i+1}> = <\beta_i^\Delta,\beta_{i+1}^\Delta> = s_j \in \mathcal{S}$. With reference to Figure 2, set $I_\eta = \{0,1,...,j\} \cup I_3$, where I_3 consists of the indices of $<\beta_{i-1},\beta_i>$, $<\beta_{i+1},\beta_{i+2}>$, $<\beta_{i+2},\beta_{i+3}> \in \mathcal{S}$. Also, for Δ satisfying (2b) or (2c,ii), η lies in a harmonic of the Ω-interval corresponding to some $(\Delta/\mathcal{C}_\gamma)$-interval $[\nu_{i'},\nu_{i'+1}]$. Letting $s_{j'} = <\beta_{i'},\beta_{i'+1}>$, so that $<\beta_{i'}, \beta_{i'+1}>g_{j'} = <\nu_{i'},\nu_{i'+1}>$ (see Figure 2), set $J_\eta = \{0,1,...,j'\} \cup I_3$, where I_3 consists of the indices of $<\beta_{i'-1},\beta_{i'}>$, $<\beta_{i'+1},\beta_{i'+2}>$, $<\beta_{i'+2},\beta_{i'+3}> \in \mathcal{S}$. For any ω fixed by such a G_η, set $I_\omega = I_\eta$ and $J_\omega = J_\eta$.

We have dealt with all $\omega \in \hat{\Lambda}$, a dense subset of $\bar{\Omega}$. Finally, let $\omega \in \bar{\Omega}\backslash\hat{\Lambda}$ (which forces $\omega h = \omega$). If $\omega g_m = \omega$ for all $j \in I$, just set $I_\omega = \{0\} = J_\omega$. Now suppose some $\omega g_m > \omega$ ($= \omega h$). We have $\omega = \inf \Sigma^+$ for some $\Sigma^+ \subseteq \hat{\Lambda}$. If even better $\omega = \inf \Sigma_1^+$ for some $\Sigma_1^+ \subseteq \text{supp}(h)$, pick a supporting interval Δ' of h such that $\omega \leq \inf \Delta' < \omega g_m$. For s_j the $\Delta \in \mathcal{S}$ of which Δ' is a harmonic (or in case (2b′) or (2c,ii′) for s_j some $<\beta_i^\Delta,\beta_{i+1}^\Delta>$ sufficiently far left in this

Δ), $\omega g_k \leq (\inf \Delta')g_k < \omega g_m$ when $k > j$. Set $I_\omega = \{0,1,...,j\} \cup \{m\}$. If ω is not the inf of such a Σ_1^+, then since g_m fixes α and thus every β fixed by G_α, $\omega = \inf \Sigma_2^+$ for some $\Sigma_2^+ \subseteq \{\sigma \in \bar{\Omega} \mid G_\tau \subset G_\sigma$ for some $\tau \in \text{supp}(h)\}$. Let Δ' be the supporting interval of h containing some τ such that $G_\tau \subset G_\sigma$ for some $\omega < \sigma < \omega g_m$, and proceed as before (obtaining $\omega g_k < \sigma g_k = \sigma < \omega g_m$). Finally, $\omega = \sup \Sigma^-$ for some $\Sigma^- \subseteq \hat{\Lambda}$, and a procedure similar to the one above (but using Σ^- instead of Σ^+) can be used to set J_ω.

At last we can apply Lemma 4.3 to guarantee that the pointwise $\sup f = \overset{p}{\vee} g_j \in G^{\because}$. Of course $\alpha f = \alpha$. By the remarks in (2′) about "$\eta g_j > \eta h$", $f \geq h$. □

References

[B1] R.N. Ball, *Topological lattice ordered groups*, Pacific J. Math. 83(1979), 1-26.

[B2] __________, *The structure of the α-completion of a lattice ordered group*, to appear in Houston J. Math.

[G] A.M.W. Glass, *Ordered permutation groups*, London Math. Soc. Lecture Note Ser. 55, Cambridge Univ. Press, London and New York, 1981.

[M] S.H. McCleary, *O-2-transitive ordered permutation groups*, Pacific J. Math. 49(1973), 425-429.

TELLING LEFT FROM RIGHT

Stephen H. McCleary
University of Georgia

Let $A = A(\mathbb{R})$ denote the lattice-ordered group of order-automorphisms of the chain $\mathbb{R}$ of real numbers. The *support* $\text{supp}(g)$ of $g \in A(\mathbb{R})$ means $\{\omega \in \mathbb{R} \mid \omega g \neq \omega\}$. Let $L = L(\mathbb{R})$ denote the ℓ-ideal of A consisting of those elements whose support lives on the left (i.e., is bounded above), let $R = R(\mathbb{R})$ be its dual, and let $B = B(\mathbb{R}) = L \cap R$.

QUESTION. Are A/L and A/R isomorphic (as ℓ-groups)?

Twenty-five years ago Charles Holland [4, p. 406] referred in passing to "the four mutually isomorphic ℓ-groups A/L, A/R, L/B, and R/B". Much of this is obvious: The Second Isomorphism Theorem tells us that A/L and R/B are isomorphic, and that A/R and L/B are isomorphic. Conjugation by the order-reversing homeomorphism $\omega f = -\omega$ yields a *group* isomorphism between A/L and A/R which reverses the order, and following f by $g \to g^{-1}$ yields a *lattice* isomorphism between A/L and A/R which reverses the group operation.

Holland pointed out a few years later it was unclear whether A/L and A/R were isomorphic as ℓ-groups. This question was posed by Glass in [2, p. 50], and the erroneous affirmative answer was reiterated by Anderson and Feil in [1, p. 143]. In his talk in Curaçao, Holland mentioned that he had continued to work intermittently on the question, and thought that a negative answer would be of more interest than a positive one. I want to thank both Holland and Manfred Droste for helpful discussions in Curaçao; in particular, Holland pointed out an error in my first "solution".

J. Martinez (ed.), Ordered Algebraic Structures, 93–97.

THEOREM. The ℓ-groups A/L and A/R are not isomorphic, or even elementarily equivalent.

There are a couple of ways to recast this result:

(1) The ℓ-groups L and R are not isomorphic (see [5, Theorem 14] or [3, Corollary 2.1.4]), and the theorem says that this remains true when B is modded out, i.e., that the ℓ-groups L/B and R/B are not isomorphic.

(2) There has been some recent interest in A^* (A with the group operation reversed, so that in A^*, gh is the result of doing h first and then g). The ℓ-groups A and A^* *are* isomorphic. (For $g \in A^*$, let $g\phi$ denote the element of A such that $\omega(g\phi)$ is that τ for which $gA_\omega g^{-1}$ (operations in A) $= A_\tau$, and follow ϕ by the above f. As usual, A_ω denotes the stabilizer subgroup $\{k \in A \mid \omega k = \omega\}$.) The theorem says that the ℓ-groups A/L and A^*/L are not isomorphic.

Proof of the theorem. Although the elements of A/L are cosets, we shall speak of an element g of A as being an element of A/L. Of course, $g_1 \leq g_2$ in A/L iff eventually $g_1 \leq g_2$ in A.

Recall how to tell when $g_1, g_2 \in A$ are conjugate. Let $g \in A$. For $\omega \in \mathbb{R}$, the supporting interval $\{\tau \in \mathbb{R} \mid \omega g^{-n} \leq \tau \leq \omega g^n$ for some $n \in \mathbb{Z}\}$ is *positive* (*negative*, *zero*) iff $\omega g > \omega$ ($\omega g < \omega$, $\omega g = \omega$). $I(g)$ denotes the chain of supporting intervals of g, the order being that inherited from $\mathbb{R}$. Then g_1 and g_2 are conjugate iff there exists an o-isomorphism between $I(g_1)$ and $I(g_2)$ which preserves the signs of the supporting intervals ([4, Lemma 10] or [3, Theorem 2.2.4]).

As elements of A/L, g_1 and g_2 are conjugate iff there exist right rays of $I(g_1)$ and $I(g_2)$ between which there exists a sign-preserving o-isomorphism.

We use this information to characterize in the ℓ-group A/L those positive elements g which are *fixed point free*, meaning that as an element of A, g has the property that there exists $\omega \in \mathbb{R}$ such that $\tau g > \tau$ for all $\tau > \omega$. Specifically, g is fixed point free iff every $h \in A/L$ is exceeded by some conjugate of g. Negative fixed point free elements are defined analogously.

In A, let t denote translation by -1 ($\omega \to \omega - 1$). Let g denote the element shown in Figure 1, which has a bump g_0 with end points at 0 and 1, a bump g_1 with end points at $1\frac{1}{2}$ and 2 (with $f^{-1}g_1f \leq g_0$), ... (in general, a bump g_n with end points at $n + 1 - \frac{1}{n+1}$ and $n + 1$, with $f^{-1}g_nf \leq g_{n-1}$).

Figure 1

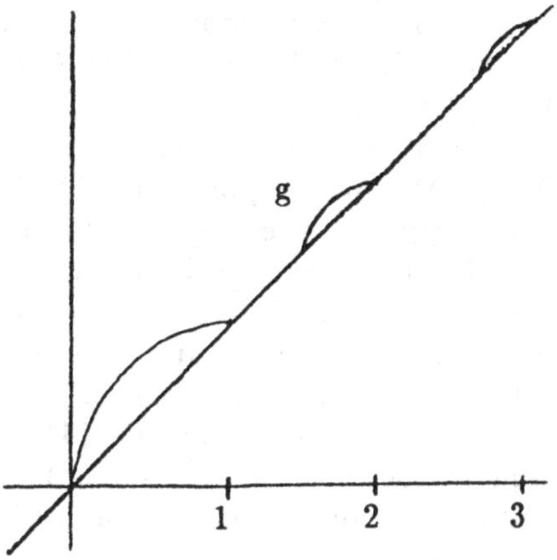

As elements of A/L, t and g have the following properties:

(1) t is negative and fixed point free.

(2) g is strictly positive.

(3) $t^{-1}gt < g$.

(4) For any square root s of t^{-1}, $g \vee s^{-1}gs$ is not fixed point free.

Property (4), which is best skipped on first reading, requires a bit of standard information about square roots. In A, any square root s of t^{-1} (i.e., of translation by +1) must work like this: For any $\omega \in \mathbb{R}$, $\omega s = \omega + \alpha$ for some $0 < \alpha < 1$, making $(\omega + \alpha)s = \omega + 1$, making $(\omega + 1)s = \omega + 1 + \alpha$,.... In particular, for all $n \in \mathbb{Z}^+$, $(\omega + n)s = \omega + n + \alpha$. Thus the map $g_n \to s^{-1}g_ns$ from the set of bumps of g onto the set of bumps of $s^{-1}gs$ must move all upper end points of bumps up by the same amount α $(0 < \alpha < 1)$. If s is any square root of t^{-1} in A/L , then as an element of A, s must eventually have the preceding property; and since eventually the bumps of g have length less than $1 - \alpha$, $g \vee s^{-1}gs$ cannot be fixed point free.

For later use, note that in A, given any $0 < \alpha < 1$, any o-isomorphism from $[0,\alpha]$ to $[\alpha,1]$ can be (uniquely) extended to a square root of t^{-1}.

To complete the proof, we show that in A/R, there can exist no pair of elements satisfying (1)-(4). Here, of course, "eventually" and "fixed point free" refer to the leftward direction. The characterization in A/R of "fixed point free" is identical to that in A/L.

Suppose by way of contradiction that $t, g \in A/R$ do satisfy (1)-(4). We may suppose that t is again translation by -1 (since all negative fixed point free elements of A/R are conjugate). By (3), the graph of g must eventually start getting greater with each leftward step of length one; i.e., eventually $g_{-(n+1)} \geq t^{-1}g_{-n}t$ where g_{-n} denotes the restriction to $[-(n+1), -n]$ of $g \in A$; see Figure 2.

Now the discussion following (4) makes it clear how to choose a square root s of t^{-1} such that the bumps of $s^{-1}gs$ together with those of g include all sufficiently small points, violating (4). □

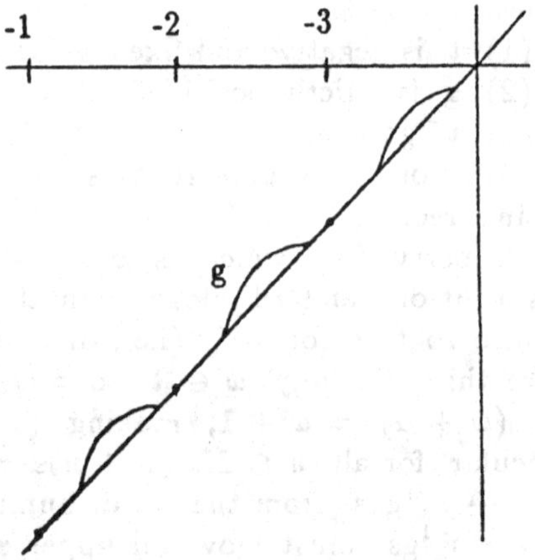

Figure 2

Of course, the above theorem remains true when $\mathbb{R}$ is replaced by an arbitrary doubly homogeneous chain of countable coterminality. Also, A/L for one such chain cannot be isomorphic (or elementarily equivalent) as an ℓ-group to A/R for any doubly homogeneous chain whatever. (If the second chain A/R had uncountable coinitiality, then A/R would contain no element satisfying the characterization of fixed point free.)

Finally, we replace one question by another:

QUESTION. Does the above theorem (either version) hold when $\mathbb{R}$ is replaced by the long line?

REFERENCES

1. M. Anderson and T. Feil, *Lattice-ordered groups: an introduction*, Reidel, Boxton, 1988.

2. A.M.W. Glass, *Ordered permutation groups*, Bowling Green State University, Bowling Green, Ohio, 1976.

3. ___________, *Ordered permutation groups*, London Math. Soc. Lecture Note Series 55, Cambridge University Press, London, 1981.

4. W.C. Holland, *The lattice-ordered group of automorphisms of an ordered set*, Michigan Math. J. 10(1963), 399-408.

5. S.H. McCleary, *The closed prime subgroups of certain ordered permutation groups*, Pacific J. Math. 31(1969), 745-753.

APPLICATIONS OF SPACES WITH FILTERS TO ARCHIMEDEAN ℓ-GROUPS WITH WEAK UNIT

RICHARD N. BALL
University of Denver
Denver, CO 80208
U.S.A.

and

ANTHONY W. HAGER
Wesleyan University
Middletown, CT 06457
U.S.A.

Abstract/Introduction

Let $\mathcal{W}$ be the category of archimedean ℓ-groups with weak unit. The usual Yosida representation of $G \in \mathcal{W}$ is one of the many embeddings of G into a $D(X)$ taking the unit to the constant function 1, but the only one with G separating the points of X. It has the further pleasure of being contravariantly functorial from $\mathcal{W}$ to compact Hausdorff spaces. We label this as $G \mapsto \hat{G} \subseteq D(YG)$, and the functor as $\mathcal{W} \xrightarrow{Y} \text{Comp}$. (The situation is described more fully in the next section.)

In a recent series of papers [BH_1--BH_4], and in further work not yet submitted, we have analyzed various aspects of the theory of epimorphisms in using the Yosida representation. The principal datum for this analysis was the filter base $\hat{G}^{-1}(R) = \{\hat{g}^{-1}(R) \mid g \in G\}$ of dense subspaces of YG. We were thus led to formalize the category SpFi of spaces equipped with a filter base of dense subsets and the SpFic-Yosida functor $\mathcal{W} \xrightarrow{SY} \text{SpFi}$ (see 8.3 of [BH_1]) and to undertake a study of SpFi itself (the first finished product of which is [BHM]).

This paper surveys some ideas in $\mathcal{W}$, mainly epimorphisms and kernels, from the explicit point of view of the grounding $\mathcal{W} \xrightarrow{SY} \text{SpFi}$. That is to say, we make definitions and constructions, and prove theorems in and about SpFi which, while natural enough topologically, are, so to speak, the images under SY of exactly what we want in $\mathcal{W}$, and then mirror the results in $\mathcal{W}$. Among the products in $\mathcal{W}$ of this process are the characterizations of epimorphisms and

J. Martinez (ed.), Ordered Algebraic Structures, 99–112.

epicompleteness from [$BH_{1,2}$], the description of the functorial epicompletion from [BH_3], and the descriptions (first presented here) of the $\mathcal{W}$-kernel-distinguishing and -preserving extensions of a $\mathcal{W}$-object.

1. The usual, and the SpFic, Yosida representation

More explicitly now, an object of $\mathcal{W}$ is an archimedean ℓ-group G with a distinguished positive weak order unit e_G, and a morphism of $\mathcal{W}$ is an ℓ-group homomorphism $G \xrightarrow{\varphi} H$ with $\varphi(e_G) = e_H$. For $G \in \mathcal{W}$, YG is the set of ℓ-ideals of G which are maximal for not containing e_G, carrying the hull-kernel topology. For X a topological space, and $[-\infty,+\infty]$ denoting the two-point compactification of the reals R, let $D(X) = \{f : X \to [-\infty,+\infty] \mid f$ is continuous, $f^{-1}(R)$ is dense$\}$. (In general, $D(X)$ need not be a group: For example, with $X = [-\infty,+\infty]$, f the obvious extension of $x + \sin x$, g the extension of $-x$, there is no h which satisfies "$h = f + g$ in $D(X)$". In §3, however, we shall encounter an important class of spaces X with $D(X)$ an ℓ-group.) Details on the following can be found in [Y], [BKW], and [HR].

1.1 THEOREM. (a) For each $G \in \mathcal{W}$, YG is compact Hausdorff, and there is a $\mathcal{W}$-isomorphism $G \to \hat{G}$ onto an ℓ-group G in $D(YG)$, with $e_G = 1$ and G separating the points of YG.

(b) If $G \to \overline{G}$ is a $\mathcal{W}$-isomorphism onto the ℓ-group $\overline{G}$ in $D(X)$, with X compact Hausdorff, $e_G = 1$ and $\overline{G}$ separating the points of X, then there is a homeomorphism $\tau : X \to YG$ with $\overline{g} = g \circ \tau$ for each $g \in G$.

(c) If $\varphi : G \to H$ is in $\mathcal{W}$, then there is a unique continuous function $Y\varphi : YH \longrightarrow YG$ with $\widehat{\varphi(g)} = \hat{g} \circ Y\varphi$ for each $g \in G$.

(d) The operator on G's given in (a), and on φ's given in (c), provides a contravariant functor $Y : \mathcal{W} \to$ Compact Hausdorff Spaces = Comp.

For $G \in \mathcal{W}$, let $G^* = \{g \in G \mid \exists\, n \ni |g| \leq n e_G\}$, and give G^* the weak (indeed, strong) unit e_G. Then $G^* \in \mathcal{W}$, and 1.1 shows that $YG^* = YG$. In the next section we describe how Y acts just on those G's with $G = G^*$.

For now, we pass directly to the SpFic Yosida representation.

1.2 DEFINITION. The category SpFi (of Spaces with Filters) has objects $(X,\mathfrak{F})$, X a compact Hausdorff space and $\mathfrak{F}$ a filter base of dense cozero subspaces (open F_σ-sets), and a morphism $\tau : (X,\mathfrak{F}) \to (Y,\mathfrak{G})$ of SpFi is a continuous map $\tau : X \to Y$ for which $\tau^{-1}(G) \in \mathfrak{F}$ for each $G \in \mathfrak{G}$.

(A caution on terminology: The present SpFi is a full subcategory of what's called ω_1-SpFi in [BHM], and differs a bit from the SpFi of [BH$_1$].)

Whenever $G \in \mathcal{W}$, we view $G \subseteq D(YG)$ (suppressing the $\wedge$ of 1.1), and have the naturally associated filter $G^{-1}(R) = \{g^{-1}(R) \mid g \in G\}$. We let $SY = (YG, G^{-1}(R)) \in$ SpFi.

1.1 (c) says that for $G \xrightarrow{\varphi} H$ in $\mathcal{W}$, the map $YG \xleftarrow{Y\varphi} YH$ satisfies $\varphi(g) = g \circ Y\varphi$, so that $\varphi(g)^{-1}(R) = (Y\varphi)^{-1}(g^{-1}(R))$ lies in $H^{-1}(R)$. We let $SY\varphi$ stand for the SpFi-Morphism $(YG, G^{-1}(R)) \leftarrow (YH, H^{-1}(R))$ which arises. Thus,

1.3 COROLLARY. The functor $Y : \mathcal{W} \to$ Comp. of 1.1 actually provides a functor $SY : \mathcal{W} \to$ SpFi.

One notes: For $G \in \mathcal{W}$ and $g \in G$, $g \in G^*$ iff $g^{-1}(R) = YG$; so $G = G^*$ iff $G^{-1}(R) = \{YG\}$.

2. The action of Y on the strong unit subcategory

Let $\mathcal{S}$ be the full subcategory of $\mathcal{W}$ of G's with $G = G^*$; this means that, in the Yosida representation, $G \subseteq C(YG)$. Let $Y^* : \mathcal{S} \to$ Comp be the restriction $Y|\mathcal{S}$. We now describe how Y^* does for $\mathcal{S}$ (and quite simply) what SY will be seen to do for $\mathcal{W}$ (at considerable labor), namely, explain epimorphism and kernels.

We set some general definitions.

2.1 DEFINITION. In a category: A morphism m is monic if $(mg = mh \Rightarrow g = h)$, and e is epic if $(ge = he \Rightarrow g = h)$. (Our convention is that mg means m acts after g.) A object G is epicomplete if $G \overset{f}{\to} \cdot$ monic and epic implies f is an isomorphism (or, more loosely, G has no proper epic extension). An epicompletion of G is a monic epic $G \overset{e}{\to} H$ with H epicomplete (or, more loosely, an epic embedding of G in an epicomplete object).

2.2 THEOREM. (a) A continuous function is Comp-monic iff it is one-to-one.
(b) $G \overset{\varphi}{\to} H$ is $\mathcal{S}$-epic iff $\varphi(G)$ separates the points of Y^*H, i.e. (via 1.1), $Y^*\varphi(G) = Y^*H$.
(c) (Thus) $G \overset{\varphi}{\to} H$ is $\mathcal{S}$-epic iff $Y^*\varphi$ is comp-monic.
(d) (Thus) an embedding $G \leq H$ is $\mathcal{S}$-epic iff $Y^*\varphi : YH \to YG$ is a homeomorphism.

2.3 COROLLARY. In $\mathcal{S}$,

(a) G is epicomplete iff $G = C(Y^*G)$,

(b) the embedding $G \leq C(Y^*G)$ is the unique epicompletion of G, and has this universal mapping property: if $\varphi : G \to C(X)$ is in $\mathcal{S}$, there is unique $\overline{\varphi} : C(Y^*G) \to C(X)$ with $\overline{\varphi} \mid G = \varphi$.

The mapping property in (b) says $C(Y^*G)$ is the "epicomplete epireflection" of G in $\mathcal{S}$.

Some of the details of 2.2 and 2.3 can be found in [L] and [Mo].

We turn to kernels.

2.4 DEFINITION. Let $G \in \mathcal{S}$. The kernel of an $\mathcal{S}$-morphism out of G is called an $\mathcal{S}$-kernel of G, and the collection of all $\mathcal{S}$-kernels of G is denoted $\mathcal{S}k(G)$. For $I \subseteq G$, let $sk_G I$ be the smallest $\mathcal{S}$-kernel which contains I (and this exists).

If I is an ideal of G, and e is a strong unit of G, then in G/I, $e \in I$ is a strong unit. Thus, $I \in \mathcal{S}k(G)$ iff G/I is archimedean. This is a nontrivial condition equivalent to "I is relatively uniformly closed"; see [LZ].

For X a space, let $K(X)$ be the complete lattice of closed subsets of X (with operations $\wedge F_\alpha = \cap F_\alpha$, $\vee F_\alpha$ = closure of $\cup F_\alpha$, with top X and bottom $\emptyset$).

2.5 PROPOSITION. Let $G \in \mathcal{S}$.

(a) $\mathcal{S}k(G)$ is a complete lattice under the operations

$$\wedge I_\alpha = \cap I_\alpha, \ \vee I_\alpha = sk_G(\cup I_\alpha)$$

with top G and bottom $\{0\}$.

(b) $k(G)$ is lattice anti-isomorphic to $K(Y^*G)$, via

$$I \mapsto F_I = \cap\{g^{-1}(0) \mid g \in I\}$$

whose inverse is $F \mapsto I_F = \{g \in G \mid g^{-1}(0) \supseteq F\}$.

(2.5 is essentially 40 of [GJ].)

Now consider an embedding $G \leq H$ with Yosida map $Y^*G \overset{\tau}{\leftarrow} Y^*H$, which is a surjection. (If the embedding were called φ, τ would be called $Y^*\varphi$.) Consider the mapping $\mathcal{S}k(H) \ni I \mapsto I \cap G \in \mathcal{S}k(G)$. (This is a convenient shorthand which means: consider the mapping $F : \mathcal{S}k(H) \to \mathcal{S}k(G)$ defined by $F(I) = I \cap G$, for $I \in \mathcal{S}k(H)$. The convention will be used several times in the sequel.) In the notation of 2.5, we have $I_F \cap G = I_{\tau(F)}$. One sees that this map of kernels is onto $\mathcal{S}k(G)$, and, evidently, is one-to-one exactly when the map $F \mapsto \tau(F)$ of closed sets is one-to-one, i.e., τ is one-to-one.

We call such an embedding $\mathcal{S}$-kernel-preserving ($\mathcal{S}$kp), and using 2.2, we have

2.6 COROLLARY. In $\mathcal{S}$,

(a) An embedding is $\mathcal{S}$kp iff it is epic,

(b) G has no proper $\mathcal{S}$kp-extension iff $G = C(Y^*G)$,

(c) $G \leq C(Y^*G)$ is the maximum $\mathcal{S}$kp-extension of G.

This is, admittedly, ado about little. The point is to set the stage for the analysis of W-kernels in §4, which is much more complicated.

3. SpFi-monics and -epics

This section is the analogue for $SY : \mathcal{W} \to \text{SpFi}$ of 2.2 and 2.3, which are about $Y^* : \mathcal{B} \to \text{Comp}$.

3.1. THEOREM. The SpFi-morphism $(Y,\mathfrak{D}) \overset{\tau}{\leftarrow} (X,\mathfrak{F})$ is monic iff: whenever $x_1 \neq x_2$ in X, there are neighborhoods U_i of x_i, and a sequence $F_1, F_2 \ldots$ from $\mathfrak{F}$ for which

$$\tau(U_1 \cap \bigcap_n F_n) \cap \tau(U_2 \cap \bigcap_n F_n) = \emptyset.$$

One might describe 3.1 as saying: The filter base $\mathfrak{F}$ hides where τ fails to be 1-1.

3.1 is a translation into SpFi of the main theorem of [BH_1] (which characterizes $\mathcal{W}$-epics). The proof is, likewise, a translation of the proof from [BH_1]. It is not especially easy. An account shall be published shortly [BH_6].

A topological space is called basically disconnected if each cozero set has open closure. Some features of these spaces are described in [GJ]. They arose originally as the Stone spaces of σ-complete Boolean algebras.

3.2 THEOREM. The SpFi-object $(X,\mathfrak{F})$ has no proper monic preimage (one might say "is monofine") iff X is basically disconnected and $\mathfrak{F}$ is the filter base of <u>all</u> dense cozeros.

3.2 differs insignificantly from the ω_1-case of the main theorem of [BHM], which, in turn, is a generalization of a theorem in Boolean Algebra [L].

We are about to construct the "maximal monic preimages" of an $(X,\mathfrak{F}) \in \text{SpFi}$, first constructing one of these with a universal mapping property. This construction, and several similar ones later, involve the Boolean σ-algebra of Baire sets in a topological space X, by definition, the smallest σ-algebra of subsets of X containing all zero-sets (equivalently, with respect to which all continuous functions to the reals become measurable). One may use [Si] as a reference for Baire sets.

The result will appear in [BH_6]. It is a topological version of the main result of [BH_3] (see 3.4(c) below), and generalizes (dually) a construction in [Si] of "σ-extensions" of Boolean algebras.

3.3 THEOREM. Let $(X,\mathfrak{F}) \in$ SpFi . Consider the σ-field B of Baire sets in X, its σ-ideal I generated by $\{Z \mid X-Z \in \mathfrak{F}\}$, the Stone representation space of the Boolean algebra $\mathcal{B}/I$ --call this space $X^\#$ --and give $X^\#$ the filter base $\mathfrak{F}^\#$ of all dense cozeros of $X^\#$, so $(X^\#,\mathfrak{F}^\#) \in$ SpFi. Then,

(a) there is a natural surjection $(X,\mathfrak{F}) \overset{\pi}{\leftarrow} (X^\#,\mathfrak{F}^\#)$ which is SpFi-monic,

(b) $(X^\#,\mathfrak{F}^\#)$ is monofine in SpFi (3.2),

(c) This universal mapping property is enjoyed: Whenever $(X,\mathfrak{F}) \overset{f}{\leftarrow} (Y,\mathfrak{G})$ is in SpFi, with $(Y,\mathfrak{G})$ monofine, there is unique $(X^\#,\mathfrak{F}^\#) \overset{\bar{f}}{\leftarrow} (Y,\mathfrak{G})$ in SpFi with $\pi\bar{f}=f$; and $\bar{f}[Y]$ is a P-set in $X^\#$. (See below.)

(d) If $(X,\mathfrak{F}) \overset{f}{\leftarrow} (Y,\mathfrak{G})$ is SpFi-monic, with $(Y,\mathfrak{G})$ monofine, then the $\bar{f}$ provided by (c) is one-to-one. Conversely, whenever Y is a P-set in $X^\#$, then $(Y, \mathfrak{F}^\#|Y)$ is monofine in SpFi, and $(X,\mathfrak{F}) \overset{\pi|Y}{\longleftarrow} (Y,\mathfrak{F}^\#|Y)$ is SpFi-monic.

(c) says that $(X^\#,\mathfrak{F}^\#)$ is the "monofine monoreflection" of $(X,\mathfrak{F})$ in SpFi.

Regarding the "P-sets" in (c) and (d) above: A subset S of a space Z is a P-set in Z if each G_δ-set containing S is a neighborhood of S. (The origins are discussed in [BHM].) The relevance here (implicit in (c) and (d) above) is essentially that, for Z basically disconnected, each dense cozero set F of Z has $F \cap S$ dense in S iff S is a P-set. (See [BHM].)

The point of this is the following paraphrase of the facts about P-sets in (c) and (d): The P-sets in $X^\#$ yield <u>all</u> the monofine monics into $(X,\mathfrak{F})$.

We now apply 3.1--3.3 to images under the functor $SY : \mathcal{W} \to$ SpFi. With a little further work, though truly not much, we find:

3.4 THEOREM. (1) $G \overset{\varphi}{\to} H$ is $\mathcal{W}$-epic iff $SY\varphi: (YH,H^{-1}(R)) \to (YG,G^{-1}(R))$ is SpFi-monic, i.e., the condition in 3.1 obtains.

(2) G is $\mathcal{W}$-epicomplete iff YG is basically disconnected and $G = D(YG)$.

(3) For $G \in \mathcal{W}$, consider $(YG, G^{-1}(R)) \overset{\pi}{\leftarrow} (YG^{\#}, G^{-1}(R)^{\#})$ from 3.3. Then, $D(YG^{\#})$ is $\mathcal{W}$-epicomplete, and the map $G \overset{\beta}{\rightarrow} D(YG^{\#})$ given by $\beta(g) = g \circ \Pi$ is a $\mathcal{W}$-epic embedding with this universal mapping property : whenver $G \overset{\varphi}{\rightarrow} H$ is in $\mathcal{W}$, with H epicomplete, there is unique $D(YG^{\#}) \xrightarrow{\overline{\varphi}} H$ with $\overline{\varphi}\beta = \varphi$.

The epicompletions of G are exactly the quotients over G of $D(YG^{\#})$.

3.4 (3) says that $D(YG^{\#})$ is the "epicomplete epireflection" of G (denoted βG in [BH_3]).

The reader has probably noticed that $D(YG^{\#})$ is asserted to be an ℓ-group (in contrast to a general $D(X)$). This is because $YG^{\#}$ is basically disconnected (3.3(b) and 3.2), hence is an F-space (each cozero set is C^*-embedded), hence is a quasi-F space (each <u>dense</u> cozero set is C^*-embedded), and it is exactly that which makes $D(X)$ an ℓ-group (an observation of Henriksen and Johnson).

In the above, 3.4(i) follows from 3.1, and is the main theorem of [BH_1].

4. SpFi-subsets and -kernels

We now describe how $SY : \mathcal{W} \to SpFi$ does for the analysis of $\mathcal{W}$-kernels what $Y^* : \mathcal{S} \to comp$ does for $\mathcal{S}$-kernels, though with a lot more work.

4.1 DEFINITION. Let $G \in \mathcal{W}$. The kernel of a $\mathcal{W}$-morphism out of G is called a $\mathcal{W}$-kernel of G, and the collection of all $\mathcal{W}$-kernels of G is denoted $\mathcal{W}k(G)$. For $I \subseteq G$, the smallest $\mathcal{W}$-kernel of G which contains I is denoted $wk_G I$ (and this exists.).

Once again, an ideal is a $\mathcal{W}$-kernel iff it is relatively uniformly closed. One may note that the only $\mathcal{W}$-kernel of G containing the weak unit is G itself.

4.2 DEFINITION. Let $(X,\mathcal{F}) \in SpFi$. A closed subset $S \subseteq X$ for which $F \in \mathcal{F}$

implies $F \cap S$ is dense in S is called a SpFi-subset of $(X,\mathcal{F})$. Then, $\mathcal{F}|S = \{F \cap S \mid F \in \mathcal{F}\}$ is a filter base of dense cozeros of S, $(S,\mathcal{F}|S) \in$ SpFi, and the inclusion $(S,\mathcal{F}|S) \hookrightarrow (X, \ )$ is a SpFi-morphism. Let $\mathrm{sub}(X,\mathcal{F})$ denote the collection of all SpFi-subsets of $(X,\mathcal{F})$. For $S \subseteq X$, let S' denote the largest SpFi-subset of $(X,\mathcal{F})$ contained in S (and this exists).

4.3 PROPOSITION. (a) For $G \in \mathbf{W}$, $\mathbf{W}k(G)$ is a complete lattice (indeed, a frame) with the operations $\wedge I_\alpha = \cap I_\alpha$, $\vee I_\alpha = wk_G(\cup I_\alpha)$, with top G and bottom $\{0\}$.

(b) For $(X,\mathcal{F}) \in$ SpFi, $\mathrm{sub}(X,\mathcal{F})$ is a complete lattice (indeed, a coframe) with the operations $\wedge S_\alpha = (\cap S_\alpha)'$, $\vee S_\alpha$ = closure of $\cup S_\alpha$, with top X and bottom $\emptyset$.

(c) For $G \in \mathbf{W}$, and associated $SYG = (YG,G^{-1}(\mathbb{R})) \in$ SpFi, $\mathbf{W}kG$ and $\mathrm{sub}(YG,G^{-1}(\mathbb{R}))$ are lattice anti-isomorphic via $I \mapsto S_I = \{g^{-1}(0) \mid g \in I\}$, whose inverse is $S \mapsto I_S = \{g \in G \mid g^{-1}(0) \supseteq S\}$.

Consider an embedding $G \leq H$ in $\mathbf{W}$, with SpFic-Yosida surjection $(YG,G^{-1}(\mathbb{R})) \overset{\tau}{\leftarrow} (YG,H^{-1}(\mathbb{R}))$. We have a map $\mathbf{W}k(H) \ni I \mapsto G \cap I \in \mathbf{W}k(G)$ which, in the notation of 4.3, is given by $G \cap I_S = I_{\tau(S)}$ (it turns out), i.e., the map of kernels is coded by the map $\mathrm{sub}(YH,H^{-1}(\mathbb{R})) \ni S \mapsto \tau(S) \in \mathrm{sub}(YG,G^{-1}(\mathbb{R}))$, and one of these is one-to-one, or onto, iff the other is. Accordingly:

4.4 DEFINITION. (a) The $\mathbf{W}$-embedding $G \leq H$ is called $\mathbf{W}$-kernel distinguishing, or $\mathbf{W}kd$ (respectively, $\mathbf{W}$-kernel preserving, or $\mathbf{W}k_p$) if the map of kernels $\mathbf{W}k(H) \ni I \mapsto G \cap I \in \mathbf{W}k(G)$ is injective (resp., bijective).

(b) The SpFi-surjection $(X,\mathcal{F}) \to (Y,\mathcal{G})$ is called SpFi-set distinguishing, or Ssd (resp., SpFi-set preserving, or Ssp) if the map of subsets $\mathrm{sub}(X,\mathcal{F}) \ni S \mapsto \tau(S) \in \mathrm{sub}(Y,\mathcal{G})$ is injective (resp., bijective).

We now analyze the SSd and SSp surjections, and after that, convert to $\mathbf{W}$, to explain the $\mathbf{W}kd$ and $\mathbf{W}kp$ embeddings.

4.5 LEMMA. Let $(X,\mathcal{F}) \in$ SpFi, and let $\mathcal{F}_\delta = \{\bigcap_n F_n \mid F_1, F_2, \ldots \in \mathcal{F}\}$. Let S be a closed subset of X. Then,

$S' = \bigcap\{cl(S \cap F) \mid F \in \mathcal{F}_\delta\} = \bigcap\{Z' \mid Z$ is a zero set, $Z \supseteq S\}$.

4.6 THEOREM. The following are equivalent for the SpFi-surjection $(X,\mathcal{F}) \overset{\tau}{\to} (Y,\mathcal{G})$.

(a) τ is Ssd.

(b) For each zero set Z of X there is a zero set W of Y with $\tau^{-1}(W)' = Z'$.

(c) If $x_1 \neq x_2$ in X, there are zero sets W_i of Y with $x_i \in \operatorname{int} \tau^{-1}(W_i)$, and there is $F \in \mathcal{F}_\delta$ such that $\tau(\tau^{-1}(W_1) \cap F) \cap \tau(\tau^{-1}(W_2) \cap F) = \emptyset$.

These conditions imply that τ is SpFi-monic and irreducible. The converse fails.

In the above: "irreducible" means that the image of any proper closed set is proper, and this is equivalent to: if G is open in X, there is W open in Y with $\tau^{-1}(W) \subseteq G$ and the closures of $\tau^{-1}(W)$ and G are the same. "Sequentially irreducible" has G and W replaced by cozero sets in the last condition. The significance of these in the theory of ℓ-groups is discussed in [BH4].

4.6(b) is to be compared with these conditions (and the translation into ℓ-groups in 4.9 is to be compared with [BH4]).

Also, 4.6(c) is to be compared with 3.1.

4.7 THEOREM. $(X,\mathcal{F})$ has no proper Ssd-preimage iff X is basically disconnected and $\mathcal{F}$ is the filter base of all dense cozeros of X.

4.8 THEOREM. Let $(X,\mathcal{F}) \in$ SpFi, and let $(X,\mathcal{F}) \overset{\pi}{\leftarrow} (X^\#,\mathcal{F}^\#)$ be the monofine coreflection described in 3.3. Let $X^0 = \{cl\, \pi^{-1}(X-M) \mid M$ is a first category Baire set in $X\}$, and let $p = \pi \mid X^0$. Then, $(X,\mathcal{F}) \overset{p}{\leftarrow} (X^0, \mathcal{F}^\# | X^0)$ is Ssd and the domain has no proper Ssd-preimage. And, if $(X,\mathcal{F}) \overset{f}{\leftarrow} (Y,\mathcal{G})$ is Ssd, then there is unique $(Y, \ ) \overset{g}{\leftarrow} (X^0, \mathcal{F}^\# \mid X^0)$ in SpFi with $p = fg$.

X^0 is the "basically disconnected cover" of X discussed in [V].

4.9 COROLLARY. (a) A $\mathcal{W}$kd embedding is $\mathcal{W}$-epic, with large image. The converse fails.

(b) $G \in \mathcal{W}$ has no proper $\mathcal{W}$kd-extension iff YG is basically disconnected and $G = D(YG)$.

(c) For $G \in \mathcal{W}$, with YG^0, etc., as in 4.8: $D(YG^0)$ has no proper $\mathcal{W}$kd-extension, the map $G \ni g \mapsto g \circ p \in D(YG^0)$ is a $\mathcal{W}$kd-extension of G, and, if $G \leq H$ is a $\mathcal{W}$kd-extension of G, then $H \leq D(YG^0)$ over G.

(In view of (a), (b) is equivalent to a proposition from [BH_2], and $D(YG^0)$ is what is called λG in [BH_3]).

We turn to SSp, and $\mathcal{W}$kp.

4.10 THEOREM. The following are equivalent for the SpFi-surjection $(X,\mathcal{F}) \overset{\tau}{\to} (Y,\mathcal{G})$.

(a) τ is Ssp.

(b) If $x_1 \neq x_2$ in X, there are neighborhoods U_i of x_i, and there is $G \in \mathcal{G}_\delta$ such that $\tau(U_1 \cap \tau^{-1}(G)) \cap \tau(U_2 \cap \tau^{-1}(G)) = \emptyset$.

(c) If $\tau = \rho\sigma$ in SpFi, then ρ is SpFi-monic.

4.11 THEOREM. $(X,\mathcal{F})$ has no proper Ssp-preimage iff each member of $\mathcal{F}$ is C^*-embedded in X, and whenever E is a dense cozero containing a member of $\mathcal{F}$, then $E \in \mathcal{F}$.

4.12 THEOREM. Let $(X,\mathcal{F}) \in$ SpFi, consider the inverse limit space $X^* = \varprojlim \{\beta F \mid F \in \mathcal{F}_\delta\}$ (β denoting Stone-Čech compactification), with the projection $X \overset{\alpha}{\leftarrow} X^*$, and give X^* the filter $\mathcal{F}^* = \{G \mid G$ is a dense cozero containing some $\alpha^{-1}(F)$, $F \in \mathcal{F}\}$. Then, $(X,\mathcal{F}) \overset{\alpha}{\leftarrow} (X^*,\mathcal{F}^*)$ is Ssp, and $(X^*,\mathcal{F}^*)$ has no proper Ssp-preimage. And, the SpFi surjection $(X,\mathcal{F}) \overset{f}{\leftarrow} (Y,\mathcal{G})$ is Ssp if and only if $\alpha = fg$ for (unique) $(Y,\mathcal{G}) \overset{g}{\leftarrow} (X^*,\mathcal{F}^*)$ in SpFi.

Note that, in contrast to the various maximal preimages considered earlier, X^* has no special topological properties. (Example: Consider any $(X,\{X\})$. Then $X^* = X$.) In particular, in the next result, $(YG)^*$ has no special topological properties.

4.13. COROLLARY. (a) The $\mathcal{W}$-embedding $G \leq H$ is $\mathcal{W}$kp iff 4.10(b) obtains for the "dual" Yosida map iff $G \leq K \leq H \Rightarrow G \leq K$ is $\mathcal{W}$-epic.

(b) G has no proper $\mathcal{W}$kp-extension iff the Yosida representation yields $G = \varinjlim \{C(F) \mid F \in G^{-1}(R)\}$; that is equivalent to "G is closed under countable composition (c^3)".

(c) For $G \in \mathcal{W}$, consider the construct $(YG, G^{-1}(R)) \overset{\alpha}{\leftarrow} ((YG)^*, G^{-1}(R)^*)$ of 4.12. The two $\mathcal{W}$-objects $\varinjlim \{C(F) \mid F \in G^{-1}(R)^*\}$ and $\varinjlim \{C(F) \mid F \in G^{-1}(R)_\delta\}$ are naturally isomorphic--call "it" c^3G--and, $SYc^3G = ((YG)^*, G^{-1}(R)^*)$. The map $G \ni g \to g \circ \alpha \in c^3G$ is a $\mathcal{W}$kp-extension. And, $G \leq H$ is $\mathcal{W}$kp if and only if $H \leq c^3G$ over G.

The c^3-objects originate in [HIJ]. The extension c^3G above, qua extension to a c^3-object, comes from [AH] (see also [H]). Part of the relation between c^3 and $\mathcal{W}$-kernels was developed independently in [Mad], from the point of view of locales.

Because of 4.9(c), we have $c^3G \leq D(YG^0)$ over G. As a matter of fact, a reading of 5.2 of [H] shows that c^3G can be constructed from G element-by-element within $D(YG^0)$.

Note the last condition in 4.13(a); we call this "restrictably epic". It is pointed out in [BH_1] that not every epic in $\mathcal{W}$ is restrictably epic. It turns out that, in reasonably general algebraic structures, an extension is restrictably epic iff the extension is "algebraic" in a certain model-theoretic sense. Thus, c^3G is, in that sense, a maximum algebraic extension--or algebraic closure--of G. The full discussion (and for Arch as well) will be in [BH_5].

Other papers to come in due course will contain the details of our analysis above of SpFi-sets and $\mathcal{W}$-kernels.

References

[AH] E. Aron and A.W. Hager, Convex vector lattices and ℓ-algebras, Top. and its Applic. 12(1981), 1-10.

[BH$_1$] R.N. Ball and A.W. Hager, Characterization of epimorphisms in archimedean lattice-ordered groups and vector lattices, Proc. Bowling Green Workshop 1985, A. Glass and C. Holland, editors, to appear.

[BH$_2$] R.N. Ball and A.W. Hager, Epicompleteness in Archimedean lattice-ordered groups, Trans. Amer. Math. Soc., to appear.

[BH$_3$] R.N. Ball and A.W. Hager, Epicompletion in archimedean ℓ-groups with weak unit, J. Austral. Math. Soc., to appear.

[BH$_4$] R.N. Ball and A.W. Hager, Archimedean-kernel-distinguishing extensions of an archimedean ℓ-group with weak unit, Indian J. Math. 29(1987), 351-368.

[BH$_5$] R.N. Ball and A.W. Hager, Algebraic extensions of archimedean ℓ-groups, in preparation.

[BH$_6$] R.N. Ball and A.W. Hager, Monomorphisms in spaces with filters, in preparation.

[BHM] R.N. Ball, A.W. Hager, and A.J. Macula, An α-disconnected space has no proper monic preimage, Top. and its Applic., to appear.

[BKW] A. Bigard, K. Keimel, and S. Wolfenstein, Groupes et anneaux réticulés, Springer Lecture Notes 608, Berlin-Heidelberg-New York, 1977.

[GJ] L. Gillman and M. Jerison, Rings of Continuous Functions, van Nostrand, Princeton, 1960.

[H] A.W. Hager, Algebraic closures of ℓ-groups of continous functions, in Rings of Continuous Functions, C.E. Bull, editor, Dekker Lecture Notes 95, 165-189, Dekker, New York, 1985.

[HR] A.W. Hager and L.C. Robertson, Representing and ringifying a Riesz space, Symp. Math. 21 (1977), 411-431.

[HIJ] M. Henriksen, J.R. Isbell, and D.G. Johnson, Residue class fields of lattice-ordered algebras, Fund. Math. 50 (1961), 107-112.

[La] R. Lagrange, Amalgamation and epimorphisms in m-complete Boolean algebras. Alg. Univ. 4 (1974), 177-179.

[Lo] H. Lord, Hull operators on a category of continuous functions on Hausdorff spaces, Studia Math. 55(1976) 225-237.

[Mac] A.J. Macula, thesis, Wesleyan University, 1989.

[Mad] J.J. Madden, Lattices and frames associated with an abelian ℓ-group, to appear.

[Mo] C. Monaco, On the category of archimedean vector lattices with strong unit, Master's Thesis, Wesleyan University, 1983.

[S] S. Sikorski, Boolean Algebras (3rd Edition), Springer-Verlag, Berlin, 1969.

[V] J. Vermeer, The smallest basically disconnected preimage of a space, Top. and its Applic. 17 (1984), 217-232.

[Y] K. Yosida, On the representation of the vector lattice, Proc. Imp. Acad. Tokyo 18 (1942), 339-443.

Prime ideal and Sikorski extension theorems for some ℓ-groups

Daniel Gluschankof [1]
Dto. de Matemáticas, Facultad de Ciencias Exactas y Naturales
Universidad de Buenos Aires
1428 Buenos Aires
Argentina

ABSTRACT. In the first part it is proved that the existence of prime ℓ-ideals in any ℓ-group is equivalent to the Prime Ideal Theorem for Boolean algebras (BPI). Moreover, the result implies that Birkhoff's Representation Theorem BRT) for representable ℓ-groups is equivalent to BPI, which extends the analogous result of [FH] and [L] for f-rings. In the second part, whose main theorem is equivalent with Sikorski Extension Theorem (SET), we characterize the injective abelian ℓ-groups with strong unit.

1. EXISTENCE OF PRIME ℓ-IDEALS IN ℓ-GROUPS

For terminology and notations we refer to [BKW]. We shall work in the axiomatic frame of Zermelo-Fränkel set theory (ZF) and explicitly mention any non-constructive axiom added in the hypotheses of any theorem (e.g. BPI, SET, AC (Axiom of Choice)). It is known that in ZF, the implications AC $\Longrightarrow$ SET $\Longrightarrow$ BPI hold but BPI $\Longrightarrow$ SET does not hold and it is not known whether SET $\Longrightarrow$ AC.

1.1. THEOREM. (BPI). *Let G be an ℓ-group, H a proper ℓ-ideal, $a \in G \setminus H$. There exists a prime subgroup P such that $a \notin P$. In particular, there exists a value of a.*

PROOF. Consider the lattice filter

$\uparrow H = \{x \in G / \exists h \in H \text{ such that } h \leq x\}$. It is easy to verify that $|a|^{-1} \notin \uparrow H$; then, since BPI implies, for any distributive lattice, that

---

[1] The author was supported during the elaboration of this paper by a fellowship from the CONICET (Consejo Nacional de Investigaciones Científicas y técnicas) who also gave him a grant for the travel to Curaçao.

J. Martinez (ed.), Ordered Algebraic Structures, 113–122.

any filter which does not contain some given point can be extended to a prime one which does also not contain that point, we have that there exists a prime lattice filter F such that $\uparrow H \subseteq F$ and $|a|^{-1} \notin F$. Denoting by F^c the set theoretical complement of F in G, define the subset of G

$$F^\wedge = \bigcap \{F_y / \ y \in F^c\}$$

where $F_y = \{x \in G/ \ yx, \ yx^{-1} \in F^c\} = \{x \in G/ \ y|x| \in F^c\}$. We shall show that $F^\wedge$ is a prime subgroup not containing a.

(1) $F^\wedge \subseteq F$: Let $x \notin F$. Since $x|x| = xxvxx^{-1} = xxve \geq e \in H \subseteq \uparrow H \subseteq F$ we have that $x|x| \in F$ and, by the definition of $F^\wedge$ that $x \notin F^\wedge$.

(2) $F^\wedge$ is a subgroup: For all $y \notin F$ we have that $ye \notin F$ which implies that $e \in F^\wedge$. Suppose $x, x' \in F^\wedge$ and $y \in F^c$ then we have that yx, yx^{-1}, yx' and yx'^{-1} belong to F^c which, in turn implies that $y(xx')^{-1} = (yx'^{-1})x^{-1}, y(x'x)^{-1} = (yx^{-1})x'^{-1}$, $y(xx') = (yx)x'$ and $y(x'x) = (yx')x$ belong to F^c, that is: xx' and $x'x$ belong to $F^\wedge$.

(3) $F^\wedge$ is a sublattice: Suppose $x, x' \in F^\wedge$ and $y \in F^c$, then we have that $y(x \vee x') = yx \vee yx'$. Since F is a prime filter, F^c is a (prime) ideal. Then, since yx, $yx' \in F^c$ we have that $yx \vee yx' \in F^c$. Also we have that $y(x \vee x')^{-1} = y(x^{-1} \wedge x'^{-1}) = yx^{-1} \wedge yx'^{-1} \in F^c$.

(4) $F^\wedge$ is convex: Suppose $x, x' \in F^\wedge$, $y \in F^c$ and $x \leq z \leq x'$, then we have that $yx \leq yz \leq yx'$ and $yx'^{-1} \leq yz^{-1} \leq yx^{-1}$ which, by convexity of F^c implies that yz and yz^{-1} belong to F^c.

(5) $F^\wedge$ is prime: Let be $x, x' \in G_+$ such that $x \wedge x' \in F^\wedge$ then, for all $y \in F^c$ we have that $y(x \wedge x')$ and $y(x \wedge x')^{-1}$ belong to F^c. Since $y(x \wedge x')^{-1} = yx^{-1} \vee yx'^{-1}$ and F^c is an ideal, both yx^{-1} and yx'^{-1} belong to F^c. Using that F^c is prime and that $y(x \wedge x') = yx \wedge yx' \in F^c$ we obtain that yx or yx' belongs to F^c which in turn implies that x or x' belongs to $F^\wedge$.

(6) Consider now the set $\mathcal{P} = \{C/ \ F^\wedge \subseteq C, \ a \notin C$ and C is prime$\}$ which is totally ordered. The prime subgroup $P = \bigcup \mathcal{P}$ is a value of a ∎

1.2. COROLLARY. (BPI). *Let G be an* abelian *ℓ-group, H a proper ℓ-ideal, $a \in G \backslash H$. There exists a prime ℓ-ideal I such that $a \notin I$ and $H \subseteq I$.*

PROOF. With the same notations of 1.1, let I the ℓ-ideal generated by $F^\wedge$ and H. That is: $I = \{x/ \ \exists x_1 \in F^\wedge, \ x_2 \in H$ such that $|x| \leq |x_1| + |x_2|\}$. Since any proper ℓ-ideal which contains a prime one is prime, it suffices to show that $a \notin I$.

Suppose there exist $x_1 \in F^{\wedge}$ and $x_2 \in H$ such that $|a| \leq |x_1|+|x_2|$, then $-|a| \geq -|x_1|-|x_2|$ which implies that $-|a|+|x_1| \geq -|x_2|$. Since $-|a| \in F^c$ and $|x_1| \in F^{\wedge}$ we have, by the definition of $F^{\wedge}$ that $-|x_2| \in F^c$, which contradicts the fact that $-|x_2| \in H \subseteq F$ ■

1.3. COROLLARY. (BPI). *Let G be a representable ℓ-group, H a proper ℓ-ideal, $a \in G\backslash H$. There exists a prime ℓ-ideal F_a which is maximal with respect to the condition of not containing a.*

PROOF. With the same notations of 1.1, consider the totally ordered set of prime subgroups $\mathcal{F} = \{xF^{\wedge}x^{-1} / x \in G\}$. Since G is representable we have that $\bigcap\mathcal{F}$ is a prime ℓ-ideal which does not contain a.

Now define $F_a = \bigcup\{C / \bigcap\mathcal{F} \subseteq C,\ a \notin C$ and C is an ℓ-ideal$\}$ which is a prime ℓ-ideal maximal among those not containing a ■

Birkhoff's representation theorem for equational varieties (BRT) (see [B]) states that in a given variety, any algebra can be represented as a subdirect product of subdirectly irreducible algebras of the variety. It was recently proved in [G] that in ZF BRT is equivalent to AC. Since the non-constructive part in BRT involves, for each two given different elements of the algebra a,b, finding a maximal congruence θ such that a and b are not congruent modulo θ we can state the following.

1.4. THEOREM. *BPI implies BRT for the variety of representable ℓ-groups.*

PROOF. Let $a,b \in G$ a representable ℓ-group, $a \neq b$. Since there exists a bijection between congruences and ℓ-ideals, there exists a maximal congruence θ for the condition $(a,b) \notin \theta$ if and only if there exists a maximal ℓ-ideal M for the condition $ab^{-1} \notin M$. With the notations of 1.1 and 1.3, define $H = \{e\}$, since $a \neq b$, we have that $ab^{-1} \notin H$ and can obtain $M = F_{(ab^{-1})}$ ■

Since BRT implies the existence of such congruences we shall show that, in ZF, BRT for representable ℓ-groups is actually equivalent with BPI.

1.5. THEOREM. *In ZF are equivalent BPI and the statement "in any hyperarchimedean ℓ-group any proper ℓ-ideal can be extended to a prime one".*

PROOF. Since any hyperarchimedean ℓ-group is abelian, the first implication results from 1.2.

For the reciprocal, we recall the fact that, in ZF, BPI is equivalent to the statement "in any powerset any non-principal ideal can be extended to a prime one". Then, let A be any set, I a proper non-principal ideal on $\mathcal{P}(A)$. Consider the hyper-archimedean ℓ-group $G = \{x \in \mathbb{R}^A / \ |x(A)| < \omega\}$. We can think of $\mathcal{P}(A)$ as the subset 2^A of G. Consider now the ℓ-ideal I' generated by I in G (i.e. $I' = \{x / |x| \le \sum x_i$ for some $x_1, \ldots, x_n \in I\}$). Defining $\mathrm{supp}(x) = \{a \in A / \ x(a) \ne 0\}$, we have that $\mathrm{supp}(x) \subseteq \mathrm{supp}\sum x_i \subseteq \bigcup \mathrm{supp}(x_i) \ne A$ because I is a proper ideal, then, since $\mathrm{supp}(1) = A$, we have that I' is proper in G. Using the hypothesis, there exists a prime ℓ-ideal P' which extends I'. Now define $P = I' \cap 2^A$ which is an ideal on 2^A which extends I. To conclude, let's see that it is prime: let be $x, y \in 2^A$ such that $x \wedge y \in P \subseteq P'$ which is prime, then we conclude that $x \in P$ or $y \in P$ which implies that P is also prime ∎

Having proved the last result we can conclude:

1.6. COROLLARY. *BPI is equivalent to BRT for the variety of representable ℓ-groups.*

In [FH] and [L] it was proved that BPI is equivalent to BRT for the variety of f-rings. Since any abelian ℓ-group is representable and can be also thought as an f-ring with the trivial product, we have that the two results intersect for the variety of abelian ℓ-groups.

To complete this first section we shall state two related results.

1.7. COROLLARY. (BPI). i) *Any hyper-archimedean ℓ-group is isomorphic to a subdirect product of subgroups of $\mathbb{R}$;*

ii) *Any archimedean ℓ-group with strong unit is isomorphic to a subdirect product of subgroups of $\mathbb{R}$.*

Observe that both results are well known but relying, for their proofs, on Zorn,s Lemma, equivalent to AC (see [BKW]).

For the last result of the section, let's recall a categorical definition:

For a category $\mathcal{C}$, an object A is *injective* if for any objects B and C such that B is a subobject of C and an arrow $f \in \mathrm{Hom}[B, A]$ there exists an arrow $\bar{f} \in \mathrm{Hom}[C, A]$ which extends f.

1.8. THEOREM. *BPI is equivalent to the statement that $\mathbb{R}$ is injective in* *the category of hyper-archimedean ℓ-groups and ℓ-homomorphisms.*

PROOF. This results from the facts that in any ℓ-group G an ℓ-ideal I is prime if and only if G/I is totally ordered, that all ℓ-homomorphic images of hyperarchimedean ℓ-groups are archimedean, and that all archimedean totally ordered groups are isomorphic to subgroups of $\mathbb{R}$ and from theorem 1.5 ∎

2. INJECTIVE ABELIAN ℓ-GROUPS WITH STRONG UNIT

In the sequel we shall work with abelian ℓ-groups with strong unit. We shall consider the language of ℓ-groups enriched by the constant symbol u which shall represent the strong unit. We shall denote one of those groups by $G(u)$, pointing at the strong unit. The homomorphisms shall be ℓ-homomorphisms such that map the strong unit in the strong unit. We shall work in $\mathcal{LGU}$, the category given by these groups and homomorphisms and prove that the theorems of existence of prime and maximal ℓ-ideals and of characterization of injective objects in the category are, in fact, equivalent to BPI and SET, respectively.

2.1 THEOREM. *In ZF the following statements are equivalent:*

a) *BPI;*

b) *In any group in $\mathcal{LGU}$ there exist prime ℓ-ideals;*

c) *In any group in $\mathcal{LGU}$ there exist maximal ℓ-ideals;*

d) *$\mathbb{R}(1)$ is injective in $\mathcal{LGU}$.*

PROOF. We only have to look at the fact that any value of the strong unit is a maximal ℓ-ideal and apply theorem 1.1 for the implications a) ⟹ b),c) and d). The equivalence of b),c) and d) are routine. For the implication b) ⟹ a) we can repeat the proof of theorem 1.5 looking at the fact that the group G defined in that proof has $\mathbf{1}$ as a strong unit ∎

Given a family $(G_i(u_i))_{i\in I} \subseteq Ob(\mathcal{LGU})$, and an ℓ-subgroup G of ΠG_i such that $(u_i)_{i\in I} \in G$, we denote by $G^*(u_i)_{i\in I}$ the ℓ-group with strong unit $\{g \in G \ /\ \exists n \in \mathbb{N}$ such that $|g_i| \leq nu_i$ for all $i \in I\}$ and it is easy to prove that $(\Pi G_i)^*(u_i)_{i\in I}$ is the product of the family $(G_i(u_i))_{i\in I}$ in the category $\mathcal{LGU}$. Making an abuse of notation, if G is an ℓ-subgroup of $\mathbb{R}^I$ for some index set I, we shall denote by $G^*(\mathbf{1})$ the object of $\mathcal{LGU}$ with underlying set $\{g \in G \ /\ \exists n \in \mathbb{N}$ such that $|g_i| \leq n$ for all $i \in I\}$.

2.2. COROLLARY. (BPI). *In $\mathcal{LGU}$ the products of copies of $\mathbb{R}(1)$ (in the sense of the above stated remark) are injective objects.*

For X a compact topological space, let G be an ℓ-subgroup of $D(X)$ such that $\mathbf{1}$ (the constant map to 1) belongs to G. Consider the interval $[0,\mathbf{1}] \subseteq G$. It can be given a De Morgan algebra structure defining $\neg x = \mathbf{1}-x$ and conserving the meet, join, 0 and $\mathbf{1}$ of G.

Now, define $\mathcal{B}(G) = \{x \ / \ x \in [0,\mathbf{1}]$ such that $x \vee \neg x = \mathbf{1}$ and $x \wedge \neg x = 0\}$ which is a Boolean algebra. In a more general setting, if G has a weak unit e, we can define $\mathcal{B}(G)$ in the same way, changing $\mathbf{1}$ by e, but in this case the structure of $\mathcal{B}(G)$ depends on the weak unit chosen. However, if G has a strong unit, all $\mathcal{B}(G)$ are isomorphic no matter which strong unit is chosen.

2.3. LEMMA. (BPI). *Let X be a Boolean space, then X is homeomorphic to the Stone space* $\mathcal{S}p(\mathcal{B}(D(X)))$.

PROOF. We have that $\mathcal{B}(D(X)) =$
$= \mathcal{B}(\mathcal{C}(X)) = \{g \in \mathcal{D}(X) \ / \ 0 \le g \le \mathbf{1}$ and $g \vee (\mathbf{1}-g) = \mathbf{1}$ and $g \wedge (\mathbf{1}-g) = 0\}$ which are exactly the continuous maps on X into $2 = \{0,1\}$; that is, $(\mathcal{C}(X,2))$. And since the Stone representation theorem (equivalent to BPI) states that X is homeomorphic to $\mathcal{S}p(\mathcal{C}(X),2))$, our statement is proved ∎

For a given compact space X we shall denote $\mathcal{C}(X,\mathbb{R}_d)$ the ℓ-group of all continuous maps on X with values on the real line with the discrete topology. Observe that $\mathcal{C}(X,\mathbb{R}_d)(\mathbf{1})$ and $\mathcal{C}(X,\mathbb{R}_d)^*(\mathbf{1})$ denote the same object in $\mathcal{LGU}$.

The following sequence of lemmas and corollaries shall be stated without proof because they are either well known or do not make use of BPI. Using corollary 1.7.ii), we shall feel free to consider any archimedean ℓ-group with strong unit as an ℓ-subgroup of products of $\mathbb{R}(1)$ (in $\mathcal{LGU}$).

2.4. LEMMA. *Let $G \subseteq \mathbb{R}^A$ be a complete ℓ-group such that $\mathbf{1} \in G$. The Boolean algebra $\mathcal{B}(G)$ is also complete.*

2.5. LEMMA. (BPI). *Let $G \subseteq \mathbb{R}^A$ be a divisible and complete ℓ-group such that $\mathbf{1} \in G$. Then $\mathcal{C}(\mathcal{S}p(\mathcal{B}(G)),\mathbb{R}_d)$ is isomorphic to an ℓ-subgroup of G.*

2.6. LEMMA. (BPI). *For any set A the ℓ-group $\mathbb{R}^A$ is isomorphic to* $D(\mathcal{S}p(2^A))$.

2.7. LEMMA. (BPI). *Let $G(\mathbf{1})$ be a complete and divisible ℓ-group with* strong unit, $B \subseteq G$ *a Boolean algebra which is subalgebra of* $\mathcal{B}(G(1))$. *If there exists* $f \in G$ *such that* $f \notin D(\mathcal{S}p(B))$ then $B \neq \mathcal{B}(G)$.

2.8. COROLLARY. (BPI). *In the conditions of the above stated lemma,* G *is isomorphic to* $D(\mathcal{S}p(\mathcal{B}(G)))^*(1) \simeq \mathcal{C}(\mathcal{S}p(\mathcal{B}(G)))^*(1)$.

Observe that, for a complete ℓ-group $G \subseteq \mathbb{R}^A$ such that $1 \in G$ are equivalent the properties of being divisible and having all the constant maps. Then we can state the following:

2.9. COROLLARY. (BPI). (Theorem of Stone-Weierstrass for $\mathcal{LGU}$). If $G(1)$ is complete and has all the constant maps then it is isomorphic to $\mathcal{C}(\mathcal{S}p(\mathcal{B}(G)))^*(1)$.

2.10. COROLLARY. (BPI). *For any* ℓ-*group with strong unit* $G(1)$, *it is complete and divisible (has all the constants) if and only if it is isomorphic to* $\mathcal{C}(X)^*(1)$ *for* X *extremally disconnected.*

2.11. LEMMA. (BPI). *The injective objects in* $\mathcal{LGU}$ *are archimedean.*

PROOF. Let $G(u)$ be injective in $\mathcal{LGU}$; by BPI it can be thought as an ℓ-subgroup of $\prod_I L_i(u_i)$ for some family of totally ordered groups in $\mathcal{LGU}$. For each L_i consider the product $L_i \times \mathbb{Z}$ endowed with the order given by the cone $K = \{(a,z) \,/\, a>0$ if a is archimedean equivalent to u_i or if $|a| \ll u_i$ then $z>0$, or if $|a| \ll u_i$ and $z=0$ then $a>0\}$. The ordered group L'_i constructed is totally ordered and has the element $(u_i, 0)$ as a strong unit. Observe also, that, for each $i \in I$, the element $(0,1)$ is greater than all infinitesimals (w.r.t. u_i) in L_i and, for L_i archimedean, L'_i is just the lexicographic product of L_i by $\mathbb{Z}$. It is easy to verify that, for the canonical embedding $L_i(u_i) \hookrightarrow L'_i(u'_i)$, u_i is mapped in u'_i, then we shall feel free to identify both strong units. Call $A(u)$ the ℓ-group $\prod_I L'_i(u'_i)$. Since $G(u)$ is injective and isomorphic to an ℓ-subgroup of $A(u)$, there exists a retraction $f{:}\,A(u) \longrightarrow G(u)$. Call x the element $((0,1))_{i \in I} \in A(u)$, it is easy to verify that $f(x) = 0$. Calling P'_i the ℓ-ideal consisting of all infinitesimals of L'_i, we have that the ℓ-ideal $\prod_I P'_i$ of $A(u)$ is contained in the kernel of f, which implies that $G(u)$ can be considered as a retract of the product of archimedean totally ordered objects in $\mathcal{LGU}$, $\prod_I (L_i(u_i)/P_i)$ (where P_i has the obvious sense). Then $G(u)$ is isomorphic to an ℓ-subgroup of $(\mathbb{R}^I)^*(1)$ and then, archimedean ∎

For the last part we shall recall the statement of SET (see [S]): *Let B be a boolean subalgebra of a boolean algebra B' and A a complete boolean algebra. Any homomorphism $f:B \longrightarrow A$ can be extended to all of B'.*

Now we are ready to prove our main result of injectivity in $\mathcal{LGU}$:

2.12. THEOREM. *In ZF the following statements are equivalent:*
a) SET;
b) Complete and divisible groups are the injective objects in $\mathcal{LGU}$.

PROOF. For the implication a) ⟹ b): First, by lemma 2.11 any injective object in $\mathcal{LGU}$ is archimedean; since SET implies BI which in turn implies corollary 1.7. ii), we have that any archimedean G (in particular any injective) in $\mathcal{LGU}$ can be embedded in $\mathbb{R}^I$ for a suitable set I, in particular, G is isomorphic to an ℓ-subgroup of $(\mathbb{R}^I)^*(1)$, which is complete and divisible; since any injective object is a retract of its extensions and retractions preserve divisibility and completeness, we conclude that all injective objects in $\mathcal{LGU}$ are complete and divisible.

Let be $G(u), H(u), A(v) \in Ob(\mathcal{LGU})$ where $H(u)$ is an ℓ-subgroup of $G(u)$ and $A(v)$ is divisible and complete. And suppose that $f:H(u) \longrightarrow A(v)$ is a homomorphism in $\mathcal{LGU}$. First, suppose that $A(v) \simeq \mathbb{R}(1)$. Since the kernel of f is a maximal ℓ-ideal of $H(u)$, by BPI we can extend it to a maximal ℓ-ideal M of $G(u)$ which defines an extension of f. If $A(v) \simeq (\mathbb{R}^I)^*(1)$ for some set I, we can repeat the procedure pointwise. For the general case we have that, by corollary 2.10 (using BPI, implied by SET) and the observation in the first part of this proof $A(v)$ is isomorphic to $\mathcal{C}(X)^*(1)$ for some X extremally disconnected, which can be embedded in $(\mathbb{R}^I)^*(1)$ for some set I. Then we have the following diagram:

$$\begin{array}{lcl} H(u) & \xrightarrow{\ f\ } & A(v) \simeq \mathcal{C}(X)^*(1) \hookrightarrow (\mathbb{R}^I)^*(1) \simeq \mathcal{C}(\mathcal{S}p(2^I))^*(1) \\ \downarrow & & \nearrow \\ G(u) & \xrightarrow{\qquad \tilde{f} \qquad} & \end{array}$$

Since SET implies that, for X extremally disconnected, there exists a section $X \longrightarrow \mathcal{S}p(2^I)$, we have, by duality, that there exists a retraction

$\mathscr{C}(\mathscr{P}\rho(2^I))^*(\mathbf{1}) \longrightarrow \mathscr{C}(X)^*(\mathbf{1})$ which implies that the above stated diagram can be reduced to

$$\begin{array}{ccc} H(u) & \xrightarrow[f]{} & A(v) \simeq \mathscr{C}(\mathscr{P}\rho(2^I))^*(\mathbf{1}) \\ \downarrow & \nearrow^{\tilde{f}} & \\ G(u) & & \end{array}$$

proving our claim.

For the reciprocal, observe first that the statement b) implies (in ZF) BPI: We have only to restate theorem 1.5. Let $G(\mathbf{1})$ be the same of that theorem, $H(\mathbf{1}) = \{g \in G/\ g \text{ is constant}\} \simeq \mathbb{R}(1)$ and $A(v) = \mathbb{R}(1)$. Let f be the isomorphism $H(\mathbf{1}) \simeq \mathbb{R}(1)$, then the kernel of $\tilde{f}$ is a maximal ℓ-ideal of $G(\mathbf{1})$ which induces a prime ideal of 2^A.

Now, let B and B' be boolean algebras such that B is a subalgebra of B'. Let $f: B \longrightarrow A$ be a homomorphism into a complete boolean algebra. Using Stone representation theorem (equivalent to BPI), the following diagram is induced:

$$\begin{array}{ccc} \mathscr{C}(\mathscr{P}\rho(B), \mathbb{R}_d)(\mathbf{1}) & \xrightarrow{f'} & \mathscr{C}(\mathscr{P}\rho(A))^*(\mathbf{1}) \\ i \downarrow & & \\ \mathscr{C}(\mathscr{P}\rho(B'), \mathbb{R}_d)(\mathbf{1}) & & \end{array}$$

where i is a monomorphism.

Since we are under the hypotheses of b), there exists an extension $\tilde{f}'$ of f' defined on the whole of $\mathscr{C}(\mathscr{P}\rho(B'), \mathbb{R}_d)(\mathbf{1})$. It is easy to verify that $\tilde{f} = \tilde{f}'_{|B'}$ is an extension of f ∎

REFERENCES

[B] Birkhoff, G., *Subdirect unions in universal algebra*, Bull. Am. Math. Soc., 50 (1944), 764-768.

[BKW] Bigard, A., K. Keimel and S. Wolfenstein, *Groupes et anneaux réticulés*. Springer LNM 608, New York (1977).

[FH] Feldman, D. and M. Henriksen, *f-rings, subdirect products of totally ordered rings, and the prime ideal theorem*, Proc. Koninklijke Nederlandse Akademie van Wetenschappen, Series A 91 (2), (1988), 121-126.

[G] Grätzer,G., *Birkhoff's representation theorem is equivalent to the Axiom of Choice*, Algebra Universalis, 23 1 (1986), 58-60.

[L] Luxemburg,W.A.J., *A remark on a paper by D.Feldman and M. Henriksen concerning the definition of f-rings*, Proc. Kon.Ned. Akad. Wet., Series A 91 (2), (1988), 127-130.

[S] Sikorski,R., *A theorem on extension of homomorphisms*, Ann.Soc. Pol.Math. 21 (1948), 332-335.

SOME APPLICATIONS OF DEFINABLE SPINE ANALYSIS IN ORDERED ABELIAN GROUPS

F. Lucas
Université d'Angers et U.A.753 C.N.R.S. France.

INTRODUCTION

The quantifier elimination theorem for ordered abelian groups (o.a.g.) proved by Peter Schmitt [S], reduces the test for elementary equivalence of two ordered abelian groups G and H to countably many tests for elementary equivalence of $Sp_n(G)$ and $Sp_n(H)$ (definable spine) of convex subgroups of G and H.

Using this, a model theory of ordered abelian groups can be developed; here we talk about two applications given by F. DELON and myself in [D.L.].

ELEMENTARY EQUIVALENCE AND SUBSTRUCTURE

A and B are elementarily equivalent ($A \equiv B$) iff they satisfies the same closed formulas and A is an elementary substructure of B ($A < B$) iff they satisfies the same closed formulas with parameters in A.

1) The case of abelian groups: here the language is $L = <=,+>$.
a) All divisible abelian groups are elementarily equivalent.
b) For each abelian group G, there exists integers λ, γ_p, $\alpha_{p,k}$, β_p such that

$$G \equiv Q^{\lambda} \oplus (\bigoplus_p (Z_{p^\infty})^{\gamma_p}) \oplus (\bigoplus_p (\frac{Z}{p^k Z})^{\alpha_{p,k}})) \oplus (\bigoplus_p (Z_p)^{\beta_p})$$

J. Martinez (ed.), Ordered Algebraic Structures, 123–128.

where p is a prime and :

$\lambda=0$ or $\lambda=1$

$Z_{p^\infty} = \{ x \in C^*, \exists k\ p^k x=1\}$ and

$Z_p = \{\frac{r}{s}, r,s \in Z,\ s\neq 0,\ (s,p)=1\}$

Those numbers λ, γ_p, $\alpha_{p,k}$, β_p are not uniquely determined; for example:

$$H\equiv \bigoplus_{i\in N}\left(\frac{Z}{p^iZ}\right) \equiv H\oplus Q \equiv H\oplus Z_p \equiv H\oplus Z_{p^\infty},$$

but we have invariants for elementary equivalence:

$$\lambda;\ \alpha_{p,k};\ \beta_p+\sum_{k=1}^{\lambda}\alpha_{p,k};\ \gamma_p+\sum_{k=1}^{\lambda}\alpha_{p,k}.$$

c) Each complete theory of abelian groups in the langage $L\cup\{(D_n)_{n\in N}\}$ where D_n is the n-divisibility, admits the elimination of quantifiers .

2) The case of ordered abelian groups:

a) Regularly ordered abelian groups:
An ordered abelian group G is said n-regular ($n\in N$) if for each nontrivial convex subgroup C of G, G/C is n-divisible ,and G is said to be regular if for all n it is n-regular .

i) If G is regular then there exists H s.t. H is archimedean and $H\equiv G$

ii) $H\equiv G$ iff as groups $H\equiv G$ and G and H are together dense or discrete .

b) Ordered abelian groups:

i) Elementary equivalence .

<u>Proposition</u> [P. SCHMITT] $H\equiv G$ iff for all n $Sp_n(H)\equiv Sp_n(G)$.

We have to define $Sp_n(G)$: for all $g \in G$:

B(g) is the smallest convex subgroup C of G, such that $g\in C$;

for each $n \in N$:

$A_n(g)=\cap\{C \text{ convex; } B(g)/C \text{ is n-regular}\}$,

$A_n(0) = \emptyset$,

$B_n(g)= \cup\{C \text{ convex ; } C/An(g) \text{ is n-regular}\}$.

$(A_n(g), B_n(g))$ is the n-regular jump of g, $C_n(g) = B_n(g)/A_n(g)$ is n-regular;

$F_n(g) = \cap\{ A_n(g+nh), h \in g \}$ is the greatest convex subgroup of G whose intersection with g+nG is empty.

$$\Gamma_n(g) = \frac{\{g';F_n(g) \supset F_n(g')\}}{\{g';F_n(g')\supseteq F_n(g)\}}$$

(this is a group but not an ordered group).

$Sp_n(g) = (\{A_n(g), g \in G\} \cup \{F_n(g), g \in G\}, A, F, c, P_i, Q_i)$

where A,F,c,P_i,Q_i are predicates, one for the A_n, one for the F_n, the others talking about the invariants of C_n and Γ_n.

ii) Elementary substructure .

<u>Proposition</u> [P. SCHMITT] H<G iff for each n there exists an elementary embedding i_n of Sp_n (H) into $Sp_n(G)$ s.t. $\forall h \in H$ $i_n(A(h,H))=A_n(h,G)$, $i_n(F_n(h,H))=F_n(h,G)$, and i_n preserves some predicates M(n,k), E(n,k), D(p,r,i).
(For example, M(n,k)(g) iff $B_n(g)/A_n(g)$ is discrete of first positve element f, and g=k.f).

iii) Quantifier elimination .

<u>Proposition</u> [P. SCHMITT] For each formula $\Psi(x_1...x_m)$ of the language of ordered groups there exists :
$n \in N$,
a formula φ of the language of colored chains,
terms in the language of groups with vaiables $x_1...x_m$,
and a quantifier free formula φ' in the language of groups with predicates M(n,k), E(n,k), D(p,r,i),
such that for all $g_1...g_n$ in G, G satisfies $\varphi(g_1 \cdots g_m)$ iff G satisfies $\varphi'(g_1...g_m)$ and $Sp_n(G)$ satsfies $\varphi(A_n(t_i(g_1...g_n)...,F_n(t_i(g_1...g_n))...)$.

CONVEX CLOSURE AND ELEMENTARY SUBSTRUCTURE

C_i are colored chains (linearly ordered sets):

Def. C_3 is the initial convex closure of C_1 in C_2 iff C_3 is an initial segment and $\forall c' \in C_3 \; \exists c \in C_1 \; c' < c$. C_4 is the relative final closure of C_1 in C_2 iff C_4 is a final segment and $\forall c' \in C_4 \; \exists c \in C_1 \; c > c'$.

Proposition [RUBIN] If $C_1 < C_2$ then $C_1 < C_3 < C_2$ and $C_1 < C_4 < C_2$.

Theorem 1 : If $H < G$ and F is the convex closure of H in G then $H < F < G$.

Theorem 2 : If $H < G$ and H_o is the greatest convex subgroup of G such that $H_o \cap H = \{0\}$, and if, for each n neither $\{0\}$ nor H_o is an F_n in $Sp_n(G)$, then $H < G/H_o$.

ELEMENTARY EQUIVALENCE AND PRODUCTS

First an example of the lemmas used to prove Theorems 3 and 4:

Lemma : The convex subgroups of GxH are of the form $\{0\} x C$ where C is a convex subgroup of H, and CxH where C is a convex subgroup of G.

For the definable spine of GxH we have: when $(x,y) \in$ GxH and $x \neq 0$:

a) if $A_n(x) \neq \{0\}$ then $A_n(x,y) = A_n(x) x H$;

b) if $A_n(x)=0$, either B(x) is not n divisible ,then $A_n(x,y)=\{0\}xH$, or it is n divisible, then one more,we have two cases :

i) if $Sp_n(G)$ has no greatest element $A_n(x,y)=\{0\}xH$;

ii) if it has a greatest element : $A_n(z)$, $A_n(x,y)=\{0\}xA_n(z)$.

Lexicographic products and Hahn products of ordered abelian groups are examples of the generalized products introduced by [F.V.] therefore if $G \equiv G'$ and $H \equiv H'$ then $GxH \equiv G'xH'$ and $G^m \equiv G'^m$.

The definable spine analysis enables us to give new results:

Theorem 3 : If I and J are totally ordered sets such that $I \equiv J$ then $G^I \equiv G^J$.

Theorem 4 : If G and H are o.a.g., $m \in N$ and $G^m \equiv H^m$ then $G \equiv H$.

(For an alternate approach, additional results and counter examples in the case of isomorphism, see [G] and [O]).

BIBLIOGRAPHY

[D.L.] F. Delon et F. Lucas, Inclusions et produits de groupes abéliens ordonnés étudiés au premier ordre (à paraître, Journal of Symbolic Logic)

[F.V.] S. Feferman et R. Vaught, The first order properties of products of algebraic systems, Fund. Math. XLVIII (1959), pp. 57-103.

[G] M. Giraudet, Cancellation and absorption of lexicographic powers of totally ordered abelian groups, Séminaire Structures algébriques ordonnées Paris VII 1986-1987(à paraître , Order).

[O] F. Oger, An example of two non isomorphic countable ordered abelian groups with isomorphic lexicographic squares, séminaire Structures Algébriques Ordonnées Paris VII 1986-1987(à paraître, Journal of Algebra).

[Sl] P. Schmitt, Model theory of ordered abelian groups, Habilitationsschrift, Heidelberg 1982.

[S2] P. Schmitt, Model and substructure complete theories of ordered abelian groups, in Models and sets, Proceedings Logic Colloquim Aachen 1983, G. Muller et M. Richtyer ed. , Springer-Verlag LNM 1103, Berlin 1984.

CHAPTER TWO:

RINGS

ORDERED ALGEBRAIC STRUCTURES IN ANALYSIS

W. A. J. LUXEMBURG
California Institute of Technology
Mathematics, 253-37
Pasadena, California 91125

1. INTRODUCTION

Real analysis is founded on the fundamental properties of the real number system. The arithmetic structures one studies in real analysis such as linear spaces and linear algebras are based on the additive structure and multiplicative structure of the reals. On the other hand, the order structure of the reals allows for comparisons of quantities expressed by inequalities. For more complex objects of real analysis such as real functions, positivity of their derivatives and integrals characterize monotonicity, another aspect of order.

In the more recent development of real analysis a more sophisticated version of an ordered structure defined by cones of "positive" elements emerged. It has its origin in Jordan's celebrated theorem characterizing the real functions that can be written as the difference of two monotone increasing (decreasing) functions. It expresses that the linear space of functions of bounded variation is generated by the cone of its non-negative elements, the increasing functions. It was F. Riesz [13], however, who for the first time at the 1928 I.C.M. Congress in Bologna pointed out this significant order theoretic aspect of Jordan's theorem. In this address, F. Riesz showed that such decomposition results hold more generally for certain spaces of continuous linear functionals in terms of positive linear functionals. To be more precise, F. Riesz showed then for the first time that the linear spaces of all continuous linear functionals defined on linear spaces of continuous real functions are generated by the cones of the positive linear functionals and so can be considered to be (partially) ordered linear spaces that have many properties in common with the order structure of the reals such as the Dedekind completeness property (every non-empty subset that is bounded above has a least upper bound).

J. Martinez (ed.), Ordered Algebraic Structures, 131–142.

In 1936, inspired by the address of F. Riesz, H. Freudenthal [5] published for the first time a set of axioms of a mathematical structure referred to nowadays as the structure of a vector lattice or Riesz space in which all the results of F. Riesz and much more could be reproduced. I may add, however, that the main result of Freudenthal is a general spectral theorem that includes the general spectral theorem for real measurable functions as well as the spectral theorem for bounded self-adjoint linear operators. At about the same time and independently similar sets of axioms were proposed by L. V. Kantorovitch [6] and S.W.P. Steen [14].

Of course, the structure introduced by Freudenthal which he referred to as "partially ordered modules" are examples of translation invariant distributive lattices. In fact, they are lattice ordered commutative groups with a module structure over the reals. In this context it is perhaps of some historical interest to point out that the general theory of lattice ordered groups has its origin in a 1942 paper by G. Birkhoff [2], the founder of lattice theory (see also [1]).

The name Riesz space, rather than vector lattice, appeared for the first time in N. Bourbaki's volume devoted to the theory of integration. In this book [4] it is clearly demonstrated in a chapter entitled "Riesz spaces" (Espaces de Riesz) the utility of the theory of positive linear functionals of vector lattices as initiated by F. Riesz in the theory of integration of real functions with respect to real-valued measures. We are following Bourbaki's example and will continue to refer to vector lattices as Riesz spaces.

In the thirties there also appeared Bochner's papers on the theory of integration of vector-valued or rather Banach space valued functions. But it was not until the late forties and early fifties that a more systematic study of vector-valued measures appeared. Particularly, in a number of fundamental papers, Dorothy Maharam-Stone [10], [11], [12] developed an integration theory for measures that take their values in spaces of real measurable functions. For such measures Maharam showed that the well-known results for scalar-valued measures such as the Carathéodory extension theorem, the Jordan-Hahn decomposition theorem and the Radon-Nikodym theorem can be reformulated for function-valued measures provided such measures are, what she called, *full-valued*. The purpose of the present paper is to show that Maharam's theory of function-valued measures of F-measures is part of the theory of positive linear operators between Riesz spaces. In this treatment we will see that full-valued measures correspond to interval preserving positive linear operators, which we have baptized "*positive linear operators with the Maharam property.*" Furthermore, as was shown by A. R. Schep and the present author [8], the Maharam property is equivalent to a stronger linearity property that best can be expressed in terms of the theory of modules of Riesz spaces over f-algebras. This aspect of the theory was developed jointly with B. de Pagter. A full account of the theory of Riesz modules and its relation to Maharam's theory of F-measures is in preparation.

2. THE MAHARAM PROPERTY

For terminology concerning the theory of Riesz spaces and f-algebras not explained in this and in the following sections we refer the reader to the following sources [7], [9], and [16].

Let L and M be two *archimedean* Riesz spaces and assume that, in addition, M is *Dedekind complete.* A subset A of L is called *order bounded* whenever there exists an element $0 < a_0 \in L$ such that for all $x \in A$, $|x| \leq a_0$. A linear transformation T of L into M is called *order bounded* if it maps *order bounded subsets* of L into *order bounded subsets* of M. A linear transformation T of L into M is called *positive* if $0 \leq f \in L$ implies $Tf \geq 0$. A celebrated theorem of F. Riesz [13], extending the Jordan decomposition theorem, states that a linear operator T is order bounded if and only if it can be written as the difference of two positive linear operators. Furthermore, the linear space over the field of real numbers **R** of all the order bounded linear transformations of L into M, partially ordered by the cone of the positive linear transformations, is a *Dedekind complete Riesz space* denoted by $\mathcal{L}_b(L,M)$.

A linear transformation $T \in \mathcal{L}_b(L,M)$ is called *order continuous* whenever $0 \leq f_\tau \in L$, $\tau \in \{\tau\}$ and $f_\tau \downarrow 0$ in L implies $\inf |T(F_\tau)| = 0$ in M, where $f_\tau \downarrow 0$ in L means that the partially ordered system $\{f_\tau\}$, $\tau \in \{\tau\}$ is directed downward (for all $\tau_1, \tau_2 \in \{\tau\}$, there exists a $\tau_3 \in \{\tau\}$ such that $f_{\tau_3} \leq f_{\tau_1}$ and $f_{\tau_3} \leq f_{\tau_2}$) and inf

$f_\tau = 0$ in L). The space for all order continuous elements of $\mathcal{L}_b(L,M)$ forms a band (polar set) and is denoted by $\mathcal{L}_n(L,M)$.

The family of measures considered by D. Maharam in [10] can be defined as follows: Let (X,Λ) and (Y,Ω) be two measurable spaces (a measurable space is defined to be an ordered pair consisting of a non-empty set and an algebra or σ-algebra of its subsets). We shall assume that Ω in a σ-algebra and that $\Omega_0 \subset \Omega$ is a special σ-ideal of Ω playing the role that sets of measure zero do in measure theory. The linear space $\mathfrak{M}(Y,\Omega,\Omega_0)$ of all Ω_0-equal real Ω-measurable functions is a Dedekind complete Riesz space. A mapping τ of Λ into $\mathfrak{M}$ is called an F-*measure* if it satisfies the following conditions: (i) $\tau(\phi) = 0$; (ii) τ is *additive*, i.e., if $A_1, \ldots, A_n \in \Lambda$ are mutually disjoint, then

$$\tau(\bigcup_{i=1}^{n} A_i) = \sum_{i=1}^{n} \tau(A_i);$$

(ii) τ is of *bounded variation*, i.e., for every $A \in \Lambda$, $\sup(\tau(B) : B \subset A \text{ and } B \in \Lambda)$ exists as an element of $\mathfrak{M}$; and (iii) τ is *countably additive*, i.e., $A_n \in \lambda$, $n = 1, 2, \cdots$ and $A_n \downarrow \phi$ implies $\inf|\tau(A_n)| = 0$ in $\mathfrak{M}$.

An F-measure extends in a unique way to a linear transformation T of the Riesz space $S(X,\Lambda)$ of the Λ-step functions (real Λ-measurable functions taking on only finitely many different values) into the Dedekind complete Riesz space $\mathfrak{M}$. From the properties of τ it follows immediately that T is an order bounded linear transformation of $S(X,\Lambda)$ into $\mathfrak{M}$, and so by the theorem of F. Riesz can be written as the difference of two positive linear transformations of S into $\mathfrak{M}$. In this context one usually assumes that the algebra Λ satisfies the countable chain condition. This translates into the statement that T is order continuous as well. As a consequence the theory of function-valued measures is part of the theory of order continuous linear transformations between Riesz spaces analogous to the fact that the classical theory of integration is part of the theory of order continuous linear functionals between Riesz spaces.

For this reason we shall now turn to those aspects of the theory of order bounded linear transformations that directly pertain to the theory of Maharam's F-measures.

We recall that a positive linear transformation $T \in \mathcal{L}_b(L,M)$ is said to have the *Maharam property whenever for all* $0 \leq f \in L$ *and for all* $0 \leq g \in M$ *satisfying* $0 \leq g \leq Tf$, *there exists an element* $0 \leq f_1 \leq f$ *such that* $Tf_1 = g$. In other words, T has the Maharam property whenever it is *interval preserving*. The reader should take note of the following fact. The corresponding notion for positive F-measures, in the sense of Maharam, in the form that for all $A \in \Lambda$ and for all $0 \leq g \in M$ such that $0 \leq g \leq \tau(A)$ there exists a Λ-measurable set $B \subset A$ such that $\tau(B) = g$, is a stronger version of the Maharam property in the sense that its extension T to $S(X,\Lambda)$ may have the Maharam property although the F-measure τ generating T may not be full-valued in the above sense. It has turned out, however that the Maharam property as defined above is strong enough to allow the development of the theory of positive linear transformations along the lines of the classical theory of integration.

We shall now explain by what we meant in the introduction by the remark that the Maharam property is equivalent with a stronger linearity property. To this end, we have to recall the result that if L is a Dedekind complete Riesz space, the order bounded linear operators $\pi \in \mathcal{L}_b(L) := \mathcal{L}_b(L,L)$ contained in the band (order complete and lattice order ideal) of $\mathcal{L}_b(L)$ generated by the identity operator I are those linear operators that leave all the bands of L invariant. Such operators π are usually referred to as *orthomorphisms* and the *band preserving property* may be rephrased in the form: for all $f \in L$, $\pi(f) \in \{f\}^{dd}$, where d denotes the disjoint or polar operator. The family of all orthomorphisms of L is denoted by Orth(L). The lattice operations of $\mathcal{L}_b(L)$ for the elements of Orth(L) are of a very simple form namely pointwise operations. By that we mean that if $\pi_1, \pi_2 \in$ Orth(L), then for all $f \in L$ we have $\inf(\pi_1, \pi_2)(f) = \inf(\pi_1(f),\pi_2(f))$ and $\sup(\pi_1,\pi_2)(f) = \sup(\pi_1(f),\pi_2(f))$. From this it follows immediately that Orth(L) is not only the band (of $\mathcal{L}_b(L)$) generated by the identity operator, but with the algebraic and the lattice structure induced by $\mathcal{L}_b(L)$, it is an f-algebra in the sense of Birkhoff and Pierce [3] and [16]. The main reason for this is that the elements of Orth(L) are local transformations in that they do not perturb the supports of the elements of L, i.e., if $\pi \in$ Orth(L) and if $f \in L$, then $g \perp f$ implies that $g \perp \pi(f)$. Typical examples of operators in analysis with the support preserving property are differential operators and multipliers. In the theory of Riesz spaces orthomorphisms are the natural generalization of the multiplier operators. To illustrate this by means of an example, consider the Riesz space $L := \mathbf{R}^n$ ($n \geq 1$), the n-dimensional real linear vector space ordered by the cone of the vectors with non-negative entries. It is well-known and easy to see that $\mathbf{R}^n$ is a Dedekind complete Riesz space. The orthomorphisms of $\mathbf{R}^n$ are obviously the linear transformations that can be represented by the diagonal matrices. For more examples see [15] and [16].

The order ideal of Orth(L) generated by the identity is called the *center* of L and is denoted by Z(L), and so $\pi \in Z(L)$ if and only if there exists a positive constant $\lambda > 0$ such that $|\pi| \leq \lambda I$. The elements of Z(L) are precisely those orthomorphisms that leave every order ideal of L invariant.

We are now in a position to state the following characterization contained in [8] of the Maharam property.

2.1 THEOREM

Let L *and* M *be Dedekind complete Riesz spaces and let* $T \in \mathcal{L}_n(L,M)$ *be an order continuous linear transformation of* L *into* M. *Then* T *has the Maharam property if and only if there exists an* f*-algebra homomorphism of* Orth(M) *into* Orth(L) *with the following properties:* $h(I) = I$, $h(Z(M)) \subset Z(L)$ *and for all* $\pi \in \text{Orth}(M)$ *we have that* $T(h(\pi)) = \pi(T)$. *Furthermore,* h *is an isomorphism on* Orth(K), *where* $K \subset M$ *is the band generated by* T(L) *in* M.

For a proof of this result we have to refer to [8], where the reader can also find the generalizations of the Hahn decomposition theorem and the Radon-Nikodym theorem. It may be of interest to point out that the Radon-Nikodym theorem for positive linear transformations between Riesz spaces is a direct consequence of the following factorization theorem.

2.2 THEOREM

Let L, M *and* T *be as in Theorem* 2.1. *Then* T *has the Maharam property if and only if for every positive linear transformation* $S \in \mathcal{L}_b(L,M)$ *satisfying* $0 \leq S \leq T$ *there exists a positive orthomorphism* $\pi \in \text{Orth}(L)$ *such that* $0 \leq \pi \leq I$ *and* $S = T(\pi)$.

These results suggest in a strong way to consider a Riesz space as a special module over its f-algebra of orthomorphisms. It is exactly those special ordered algebraic structures that we shall discuss in the following section.

3. f-*MODULES*

We recall that a linear lattice order algebra A over the reals or that a Riesz-algebra is called an f-*algebra* in the sense of Birkhoff and Peirce whenever for all a, b, c $\in$ A we have that b $\perp$ c implies ab $\perp$ c and ba $\perp$ c. It was shown by Birkhoff and Pierce that Archimedean f-algebras are automatically commutative. For a simple proof of this fact we refer to [15]. The f-modules to be defined below and that play an important role in the theory of positive linear transformations are exactly the Archimedean ones. For this reason we shall restrict the discussion to the Archimedean case only.

3.1 DEFINITION

Let A *be an Archimedean* f-*algebra with unit element* $0 < e \in A$ *and let* L *be an Archimedean Riesz space. Then* L *is called a left* f-*module over* A *whenever there exists a mapping* (a,f) $\to$ af *of* A $\times$ L *into* L *such that the following properties hold*:

i) *L is a left module over the linear algebra* A *w.r.t. the mapping* (a,f) $\to$ af *of* A $\times$ L *into* L. *In particular, for all* f $\in$ L, ef = f.

ii) *If* $0 \leq a \in A$ *and* $0 \leq f \in L$, *then* af ≥ 0.

iii) *If* $f, g \in L$ *and* f $\perp$ g (inf(|f|,|g|) = 0 in L) *and if* a $\in$ A, *then* af $\perp$ g.

As in the case of the definition of the general concent of modules, right f-modules are defined in a similar fashion.

An f-module L is of course also an Archimedean Riesz space and the module operation and scalar multiplication are related by property (i), in that if r is real and f $\in$ L, rf = (re)f.

If L is a left-f-module over the f-algebra A, then for each a $\in$ A, the mapping π_a *of L into L defined by* $\pi_a(f)$ = af, f $\in$ L, is an orthomorphism of L. The mapping h of A into Orth(L), defined by h(a) := π_a, a $\in$ A, is by (ii), an algebraic homomorphism and by (iii), h is also a positive linear mapping of A into Orth(L). The two properties combined show that h is an f-algebra homomorphism of A into Orth(L) satisfying h(e) = I. Thus the algebra A of operators of a left-module can be embedded into a subalgebra of Orth(L).

Conversely, if L is an Archimedean Riesz space and if A is an Archimedean f-algebra and if h is an f-algebra homomorphism of A into Orth(L) satisfying h(e) = I, then h defines a left f-module structure on L by means of the definition, for all $a \in A$ and for all $f \in L$, we define af := h(a)(f).

Since every unital Archimedean f-algebra A is semi-prime, i.e., its radical of nilpotent elements is zero or, equivalently, $a \in A$ and $a^2 = 0$ implies $a = 0$, it follows immediately that if L is a left-module over A, then the embedding h of A into Orth(L) is an f-algebra isomorphism (see [1], section 12.3).

In case L is an Archimedean Riesz space and so a left f-module over A := **R**, the f-algebra of the real numbers, then the embedding of A into Orth(L) is nothing but the identification of **R** with the subfield of Orth(L) consisting of the multiples of the identity.

Since the elements of Orth(L) are band preserving it follows immediately that if L is a left f-module over A, then any of its bands are left f-modules over A as well. Furthermore, any Archimedean Riesz space L is in a natural way a left f-module over any of the unital subalgebras of Orth(L). In particular, L is a left f-module over its *center* Z(L), the subalgebra of Orth(L) of the ideal preserving orthomorphisms of L. If L is a left f-module over A and if M is a Dedekind complete Riesz space, then the Riesz algebra $\mathcal{L}_b(L,M)$ of the order bounded linear transformations of L into M can be considered to be a right f-module over A. Indeed, if $T \in \mathcal{L}_b(L,M)$ and if $a \in A$ we define (Ta)(f) = T(af), $f \in L$. If L is an Archimedean Riesz space and if M is a Dedekind complete Riesz space and a left f-module over an f-algebra B, then $\mathcal{L}_b(L,M)$ can be considered to be a left f-module over B. Indeed, if $T \in \mathcal{L}_b(L,M)$ $b \in B$ we define (bT)(f) = b Tf, $f \in L$.

As a consequence we have the following fact: *If* L *is a left* f-*module over* A *and if* M *is a Dedekind complete left* f-*module over* B, *then* $\mathcal{L}_b(L,M)$ *is simultaneously a right* f-*module over* A *and a left* f- *module over* B.

We conclude this section with the remark that if L is a left f-module over A, the Dedekind completion $\hat{L}$ of L is a left f-module over A as well. This follows immediately from the well-known fact that every orthomorphism is order continuous (see [8] for a simple proof of this fact).

4. f-LINEARITY

We shall now introduce a stronger notion of linearity for order bounded transformations suggested by the Maharam property.

4.1 DEFINITION

Let L *and* M *be left* f-*modules over* A. *An order bounded linear transformation* $T \in \mathcal{L}_b(L,M)$ *is called* f-*linear whenever for all* $f \in L$ *and for all* $a \in A$ *we have* $T(af) = aT(f)$. *The family of all* f-*linear transformations of* L *into* M *will be denoted by* $\mathcal{L}_b(L,M;A)$.

We begin with the following observation.

4.2 THEOREM

If L *and* M *are left* f-*modules over* A *and if* M *is Dedekind complete, then* $\mathcal{L}_b(L,M;A)$ *is a band of* $\mathcal{L}_b(L,M)$.

PROOF. At the end of section 3 we have shown that $\mathcal{L}_b(L,M)$ is a right f-module as well as a left f-module over A. Since the mappings $T \to aT$ and $T \to Ta$, $a \in A$ are orthomorphisms of $\mathcal{L}_b(L,M)$, the mapping $T \to aT - Ta$ is an orthomorphism of $\mathcal{L}_b(L,M)$. From the order continuity property of orthomorphisms it follows that for all $a \in A$, the null-ideal $N_a := \{T \in \mathcal{L}_b(L,M) : aT - Ta = 0\}$ is a band. The final result then follows from the observation that $\mathcal{L}_b(L,M;A) = \bigcap(N_a : a \in A)$, and the proof is finished.

REMARKS. 1) If $T \in \mathcal{L}_b(L,M)$ and if $a \in A$, then in general $aT - Ta \neq 0$. 2) Since $\mathcal{L}_b(L,M;A)$ is a band of $\mathcal{L}_b(L,M)$ it is in its own right also a right as well as a left f-module over A. Similarly for $\mathcal{L}_n(L,M;A)$, the band of $\mathcal{L}_b(L,M;A)$ of all order continuous f-linear mappings.

In section 1 we defined the Maharam property for positive linear transformation. We shall now say *that an order bounded linear operator* $T \in \mathcal{L}_b(L,M)$ *has the Maharam property whenever its modulus* $|T|$ *does.* It is easy to see that $T \in \mathcal{L}_b(L,M)$ has the Maharam property if and only if T^+ and T^- have the Maharam property. More precisely we have the following result.

4.3 THEOREM

Let L *and* M *be Archimedean Riesz spaces and assume that* M *is Dedekind complete. If* L *is a left* f*-module over* $A := Z(M)$ *or any* f*-subalgebra of* Orth(M) *containing* Z(M), *then every* f*-linear transformation* $T \in \mathcal{L}_b(L,M;A)$ *has the Maharam property.*

PROOF. We have to show that the modulus $|T|$ of T has the Maharam property. But since we have shown in Theorem 3.1 that $\mathcal{L}_b(L,M;A)$ is a band of $\mathcal{L}_b(L,M)$ we may assume that T is positive, i.e., $0 \le T \in \mathcal{L}_b(L,M;A)$. Assume then that there exist elements $0 \le f \in L$ and $0 \le g \in M$ such that $0 \le g \le Tf$. Since M is Dedekind complete there exists an orthomorphism $\pi \in Z \subset A$ such that $0 \le \pi \le I$ and $g = \pi(T(f))$. Since T is f-linear it follows that $g = \pi(T(f)) = T(\pi(f))$ and from $0 \le \pi \le I$ it follows that $\pi(f) \le f$. Hence, there exists an element $f_1 := \pi(f) \le f$ such that $g = Tf_1$, and the proof is finished.

Assume now that $T \in \mathcal{L}_n(L,M)$ is order continuous and that $|T|$ has the Maharam property. Then, by Theorem 2.1, there exists an f-algebra homomorphism h of Orth(M) into Orth(L) with the properties, $h(I) = I$, $h(Z(M)) \subset Z(L)$ and for all $\pi \in$ Orth(M), we have $\pi(|T|) = |T|(\pi)$. Observe now that the f-algebra homomorphism h defines on L a left f-module structure over $A :=$ Orth(M) by defining for all $\pi \in$ Orth(M) and for all $f \in L$, $\pi \cdot f = h(\pi)(f)$. Then for all $\pi \in$ Orth(M) we have that $\pi(|T|) = |T|(h(\pi))$ which expresses that $|T|$ is f-linear. Since, by Theorem 4.1, the f-linear transformations form a band, we have that T is also f-linear, in symbols $\pi(|T|) = |T|(h(\pi))$ implies $\pi(T) = T(h(\pi))$ for all $\pi \in$ Orth(M).

Combining now all the above results we have the following characterization of the Maharam property. This completes the proof.

4.4 THEOREM

Let L *and* M *be Dedekind complete Riesz spaces and let* $T \in \mathcal{L}_n(L,M)$. *Then* T *has the Maharam property if and only if there exists a left* f*-module structure over* A := Orth(M) *of* L *such that* $T \in \mathcal{L}_n(L,M;\ \mathrm{Orth}(M))$, i.e., T *is an order continuous* f*-linear transformation.*

REFERENCES

[1] Bigard, A., K. Keimel and S. Wolfenstein, *Groupes et Anneaux Reticulés.* Lecture Notes in Mathematics 608, Berlin-Heidelberg-New York: Springer 1977.

[2] Birkhoff, G., *Lattice-ordered groups*, Annals of Math., 43 (1942), 298-331.

[3] Birkhoff, G. and R. S. Pierce, *Lattice-ordered rings*, Anais Acad. Brasil Ci., 28 (1956), 41-69.

[4] Bourbaki, N., *Eléments de Mathématique*, *Intégration*, Chap. 1, 2, 3 et 4, Paris; Hermann 1965.

[5] Freudenthal, H., *Teilweise geordnete Moduln*, Proc. Royal Acad. of Sc. Amsterdam 39 (1936), 641-651.

[6] Kantorovitch, L. V., *Linear partially ordered spaces*, Mat. Sbornik (N.S.2) 44 (1937), 121-168.

[7] Luxemburg, W.A.J., *Some Aspects of the Theory of Riesz Spaces*, The University of Arkansas Lecture Notes in Mathematics 4, Fayetteville, 1979.

[8] Luxemburg, W.A.J. and A. R. Schep, *A Radon-Nikodym type theorem for positive operators and a dual*, Indagationes Math., 81 (1978), 357-375.

[9] Luxemburg, W.A.J. and A. C. Zaanen, *Riesz Spaces I*, Amsterdam: North-Holland 1971.

[10] Maharam, D., *The representation of abstract measure functions*, Trans. Amer. Math. Soc., 65 (1949), 279-330.

[11] Maharam, D., *Decompositions of measure algebras and spaces*, Trans. Amer. Math. Soc., 69 (1950), 142-260.

[12] Maharam, D., *The representation of abstract integrals*, Trans. Amer. Math. Soc., 75 (1953), 154-184.

[13] Riesz, F., *Sur la décomposition des opérations fonctionelles linéaires*, Atti del Cong. Internaz. dei Mat., Bologna 1928, vol. 3 (1930), 142-146; Oeuvres Completes II, 1097-1102, Budapest: 1960.

[14] Steen, S.W.P., *An introduction to the theory of operators I*, Proc. London Math. Soc., (2) 41 (1936), 361-392.

[15] Zaanen, A. C., *Examples of orthomorphisms*, Journ. of Approx. Theory, 13 (1975), 192-204.

[16] Zaanen, A. C., *Riesz Spaces II*, Amsterdam, North-Holland, 1983.

RINGS OF CONTINUOUS FUNCTIONS

FROM AN ALGEBRAIC POINT OF VIEW

By

Melvin Henriksen

Harvey Mudd College

Claremont, CA 91711

U.S.A.

1. Introduction

When George Martinez invited me, close to a year ago to give a talk on rings of continuous functions at a conference on ordered algebraic systems in an exotic location, I felt honored and elated. The accompanying feeling of euphoria stayed with me until I started to think seriously about what I could say to an audience containing people thoroughly familiar with the Gillman-Jerison text [GJ1] and many of the ensuing developments since 1960 – a lot of them due to mathematicians to whom I would be talking. I began to wonder if I would be doing the equivalent of carrying coals to Curaçao in August.

With no further introspection, this article will be an incomplete survey of development in the study of algebraic aspects of rings of continuous functions since the publication of [GJ1] in 1960. As usual, let $C(X)$ denote the ring (with the usual coordinatewise operations of addition and multiplication) of continuous real-valued functions on a topological space X. Under suitable restrictions on X, the ring $C(X)$ determines X (This is the case, in particular if X is a compact (Hausdorff) space).

J. Martinez (ed.), Ordered Algebraic Structures, 143–174.

Studying the ring $C(X)$ makes one ask questions about the underlying space that would not arise if one looks at topological spaces in more classical ways. Most of the research activity since 1960 has been concerned with the underlying space X rather than the algebraic structure of th ring $C(X)$. Below I will emphasize the latter rather than the former. As L. Gillman reminded us in the title of [G], "Rings of Continuous Functions are Rings." I will try to revive interest in some old unsolved problems, and will state some new ones. I will give brief synopses of many papers and leave it to the interested reader to examine the details. Some historical remarks are made with no pretense of giving a complete history.

I begin with a summary of those parts of [GJ1] essential to understanding what follows. Despite its venerable age, this excellently written book must be read by any serious student of $C(X)$. Its results have been generalized and improved upon, and many of these improvements are recorded books such as [CN], [PW], [Wa], and [We], but [GJ1] is far from obsolete.

Section 1 consists mainly of a review of those parts of [GJ1] needed to set the stage for the sequel and to describe the nature of the subject, but some material that appeared after 1960 is included. In particular, some mention is made of R. L. Blair's work on z-embedding, and the work of A. Stone and W. Comfort on normality and local compactness not being preserved in going from X to vX is mentioned.

Section 2 is devoted in part to the problems involved in translating topological concepts into algebraic language and vice versa. The complicated nature of real-compact spaces seems to make this difficult. Attempts to obtain internal algebraic characterizations of a $C(X)$ are reviewed with particular emphasis on such an attempt by A. Hager. All such characterizations seem to rely on the representation theorem for Φ-algebras as algebras of extended real-valued functions on a compact space due to D. G. Johnson and me.

Section 3 consists mainly of a review of the efforts to classify H-fields (i.e., fields of the form $C(X)/M$, where M is a maximal ideal of $C(X)$, that contains $\Re$ properly. Through the use of model theory, set theory, and the theory of ultrafilters much more has been learned about H-fields than appeared in the Gillman-Jerison text, but much more remains to be done. Recent progress on integral domains that are homomorphic images of a $C(X)$ is also described.

In a final section, I mention briefly a few other related topics: discontinuous homomorphisms of $C(X)$ into a Banach algebra, other algebraic structures on $C(X)$, spaces of minimal prime ideals of $C(X)$, and covers and completions.

I have tried to give credit where it is due, but I refer to books and monographs in preference to original sources.

1.Pre-requisites; setting the stage.

By and large, the notation and terminology of [GJ1] will be used unless it conflicts with the by now standard notation used by workers in lattice - ordered rings, in which case the terminology of [BKW] is adopted. The reader's attention will be called to conflicts in terminology as they arise. A *Tychonoff space* is a subspace of a compact (Hausdorff) space. (In [GJ 1], such spaces are called *completely regular*). A space X is called *quasicompact* if every open cover of X has a finite subcover, and a Hausdorff quasicompact space is *compact*. X is a Tychonoff space if and only if whenever $F \subset X$ is closed and $x \notin F$, there is $f \in C(X)$ such that $f(x) = \{0\}$ and $f[F] = \{1\}$. It was known to both M.H. Stone and E. Čech that for each topological space X, there is a Tychonoff space X' such that $C(X)$ and $C(X')$ are isomorphic. (See the historical notes for Chapter 3 of [GJ1]). Since we regard topological spaces as the domain of continuous real-valued functions, it will be assumed henceforth that X denotes a Tychonoff space unless the contrary is stated explicitly.

For $f \in C(X)$ the *zeroset* of F is $Z(f) = \{x \in X : f(x) = 0\}$, and $\text{coz}\, f = X \backslash Z(f)$ is called the *cozeroset* of f. The set $\mathcal{Z}(X) = \{Z(f) : f \in C(X)\}$ is closed under finite union (since if $f_i \in C(X)$ for $1 \leq i \leq n$, then $\cup_{i=1}^n Z(f_i) = Z(f_1, f_2 \cdots f_n))$ and under countable intersection (since if $f_i \in C(X)$ for $i \geq 1, f = \sum_{i=1}^{\infty} \frac{|f|_i \wedge 1}{2^i}$, then $\cap_{i=1}^{\infty} Z(f_i) = Z(f))$. $C(X)$ is a commutative ring whose identity element is the constant function $\mathbf{1}$. Let $C^*(X)$ denote the set of bounded elements of $C(X)$ and note that $C^*(X)$ is a subring of $C(X)$. Observe that $f \in C(X)$ is invertible if and only if $Z(f) = \emptyset$, but this need not be the case for the subring C^*. For if N denotes the (discrete) space of positive integers, and $j(n) = 1/n$ for $n \in N$, then $Z(j) = \emptyset$, but j has no inverse in $C^*(N)$.

A subspace S of X is said to be *C-embedded* (resp. *C^*-embedded*) in X if the map $f \to f|_S$ is an epimorphism of $C(X)$ onto $C(S)$(resp. $C^*(X)$ onto $C^*(S)$). The following theorem, established independently by M.H. Stone and E. Čech, is essential to all that follows. See [GJ1, Chap. 6].

1.1 <u>Theorem</u> *Every Tychonoff space* X *is dense and* C^**-embedded in a compact space* β X. *Moreover* β X *is unique in the sense that if* X *is a dense,* C^**- embedded subspace of a compact space* K, *then there is a homeomorphism of* β X *onto* K *whose restriction to* X *is the identity map.*

There are three ways to construct βX each of which is more useful than the other two for some purposes. E. Čech proceeded as follows. Let $I(X) = C(X) \cap [0,1]^X$

and let I_f denote a copy of $[0,1]$ for each $f \in I(X)$. By the Tychonoff theorem [GJ1, Chapter 6], $P = \pi\{I_f : f \in I(X)\}$ is compact, and the map $e : X \to P$ given by $e(x)_f = f(x)$ is injective and bicontinuous. It follows that $\beta X = Cl_P(e[X])$ contains $e[X]$ as a dense C^*-embedded subspace and is compact.

M.H. Stone's construction is an algebraic one. Let $\mathcal{M}$ denote the family of maximal ideals of $C^*(X)$. If $M \in \mathcal{M}$, then $C^*(X)/M$ is a Dedekind complete ordered field and hence is isomorphic to the real field $\Re$. For $f \in C^*(X)$, let $h(f) = \{M \in \mathcal{M} : f \in M\}$ and let $h^c(f) = M \backslash h(f)$. Taking $\{h^c(f) : f \in C^*(X)\}$ as a base for a topology on $\mathcal{M}$ (thereby getting what is called the *Stone topology* or *hull-kernel* topology on $\mathcal{M}$) makes it into a compact space. The map that sends $x \in X$ to $M_x = \{f \in C^*(X) : f(x) = 0\}$ is a homeomorphism of X onto a dense subspace of $\mathcal{M}$, and, for each $f \in C^*(X)$, the map $\bar{f} : \mathcal{M} \to \Re$ such that $\bar{f}(M) = f + M$ is a continuous real-valued extension of f over $\mathcal{M}$. Thus $\beta X = \mathcal{M}$ has the properties announced in Theorem 1.

A collection $\mathcal{F}$ of zerosets of X such that (i) $\emptyset \notin \mathcal{F}$, (ii)$Z_1, Z_2 \in \mathcal{F}$ implies $Z_1 \cap Z_2 \in \mathcal{F}$, and (iii) $Z_1 \in \mathcal{F}, Z \in \mathcal{Z}(X)$ and $Z_1 \subset Z$ implies $Z \in \mathcal{F}$, is called a z-*filter*, and a maximal z-*filter* is called a z-*ultrafilter*, (In case X is a discrete space, the prefix "z - " is usually dropped). If $\mathcal{F}$ is a z-filter, then $I(\mathcal{F}) = \{f \in C(X) : Z(f) \in \mathcal{F}\}$ is an ideal of $C(X)$ which is maximal if and only if $\mathcal{F}$ is a z-ultrafilter.

The Stone topology on $\mathcal{M}$ induces a topology on the family of z-ultrafilters on X. If $M \in \mathcal{M}$, and $\mathcal{Z}(M) = \{Z(f) : f \in M\}$ is the corresponding z-ultrafilter on X, we take a base for a topology on the space $\mathcal{U}(X)$ of z-ultrafilters on X by taking $\{\{\mathcal{Z}(M) : f \notin M\} : f \in C^*(X)\}$ as a base for a (compact) topology. A z-filter $\mathcal{F}$ is called *fixed* or *free* according as $\cap\mathcal{F}$ is nonempty or empty, and the space X is homeomorphic to the dense subspace of $\mathcal{U}(X)$ if $x \in X$ is mapped onto the (fixed) z-filter of zerosets containing x. Similarly, an ideal I of $C(X)$ is called *fixed* or *free* according as $\cap\mathcal{Z}(I)$ is nonempty or empty. The image of X under this mapping is dense and C^*-embedded in X, so $\beta X = \mathcal{U}(X)$. This last construction is due to E. Hewitt [Hew] and L. Gillman and M. Jerison [GJ1, Chap 6]. The latter should be examined for more details and for a proof of the uniqueness of βX.

If X is a compact space, then every maximal ideal of $C(X)$ is fixed. That is, if M is a maximal ideal of $C(X)$, there is a unique $x \in X$ such that $M = M_x = \{f \in C(X) : f(x) = 0\}$ and the map $x \to M_x$ is a homeomorphism of X onto the space of maximal ideals of $C(X)$ with the hull-kernel topology. Thus the topology of X is determined by the ring $C(X)$ if X is compact; that is, if X_1 and X_2 are compact spaces and the rings $C(X_1)$ and $C(X_2)$ are isomorphic, then X_1 and X_2 are homeomorphic. In his pioneering paper [Hew], E. Hewitt extended this to a wider class of spaces as follows.

If X is any (Tychonoff) space and M is a maximal ideal of $C(X)$, then C/M is a totally ordered field containing $\Re$ which is called *real* or *hyper-real* according as $C/M = \Re$ or C/M contains $\Re$ properly. In the latter case, C/M is called an H-*field* in [ACCH], [Dow], and [Ro], and M is called a *hyper-real ideal.* If every free ideal of $C(X)$ is hyper-real, then X is said to be *realcompact.* By now, this terminology is standard, but such spaces are called Q-*spaces* in [Hew], *realcomplete* in [AS] and *Hewitt - Nachbin complete* in [We]. The space of maximal ideals of $C(X)$ is homeomorphic to βX under a map that leaves X pointwise fixed, and the subspace vX of real maximal ideals of $C(X)$ is called the *realcompactification* of X. Since the space of maximal ideals of $C(X)$ with the hull-kernel topology is homeomorphic with βX, the inclusion $X \subset vX \subset \beta X$ holds, and it is clear that an isomorphism of $C(X_1)$ onto $C(X_2)$

induces a homeomorphism of vX_2 onto vX_1. See [GJ1, Chap.8].

This lovely result seems at first to put us into a paradise where we can use ring-theoretic techniques to prove topological theorems and vice versa. The first hiss of the serpent comes from the complicated nature of the class of realcompact spaces and realcompactifications.

The class of realcompact spaces includes all Lindelöf spaces (and hence all separable metric spaces), is closed-hereditary, and productive. Thus every closed subspace of an arbitrary product of copies of $\Re$ is realcompact, and the converse holds as well; that is X is realcompact if and only if it is homeomorphic to a closed subspace of a product of real lines; see [GJ1, Chap. 12]. While this characterization of realcompactness is pretty, it is often difficult to apply; say, to answer the question of whether spaces that are metrizable must be realcompact. To answer this question, one must enter the world of axiomatic set theory.

An *Ulam measure* μ on a set S is a countably additive measure on S such that (i)$T \subset S$ implies $\mu(T) = 0$ or $\mu(T) = 1$, (ii) $x \in S$ implies $\mu(\{x\}) = 0$, and (iii)$\mu(S) = 1$. An infinite cardinal α is said to be (Ulam) *measurable* if there is an Ulam measure on a set S of cardinality α; otherwise α is said to be *nonmeasurable*. If X is a discrete space, M is a real ideal of $C(X)$, and $T \subset X$, let $\mu(T) = 1$ if $T \in \mathcal{Z}(M)$, and $\mu(T) = 0$ otherwise. Then μ is a countably additive measure on X which will be an Ulam measure if and only if M is free. It follows that a discrete space is realcompact if and only if the cardinality of X is nonmeasurable. S. Ulam showed in 1929 that any measurable cardinal has to be tremendously large in the following sense: (i) The smallest infinite cardinal $\aleph_0$ is nonmeasurable, (ii) any cardinal less than a nonmeasurable cardinal is nonmeasurable, (iii) the sum of nonmeasurably many nonmeasurable cardinals is nonmeasurable, and (iv) if α is nonmeasurable, so is 2^α.

If Zermelo - Fraenkel set theory with the axiom of choice (ZFC) is consistent, so is ZFC with the hypothesis that no cardinal is measurable; see [GJ1, Chap.12]. Despite this, the possible existence of a measurable cardinal makes the statement of many theorems about realcompactness awkward.

One of the deeper topological characterizations of realcompactness is due to T. Shirota who showed that X *is realcompact if and only if:* (i) *X admits a complete uniform structure compatible with its topology and* (ii) *every closed discrete subspace of X has nonmeasurable cardinality*; see [GJ1, Chap. 15].

The necessary but ugly addendum (ii) can be replaced with a euphemism in two ways. A discrete space X is realcompact if and only if it has nonmeasurable cardinality, so (ii) is equivalent to (ii)′: *every closed discrete subspace of X is realcompact.* Alternatively, in [Ho, p. 12], R. Hodel defines the *extent* $e(X)$ of X to be the maximum of ω and sup $\{\text{card.}D : D \subset X \text{ closed and discrete}\}$. So (ii) is equivalent to (ii)″ $e(X)$ *is nonmeasurable.*

A useful corollary is that *every metrizable (indeed, every paracompact) space of nonmeasurable extent is realcompact.* This follows also from a more general result given in [PW, 5.11 (1)].

A space X is said to be *z-embedded* in a space Y if each $Z \in \mathcal{Z}(X)$ is the intersection with X of some $Z' \in \mathcal{Z}(Y)$. Every C^*-embedded and every cozero set of a space Y is z-embedded in Y as is every Lindelöf space contained in it. The notion of z-embedding was introduced by R. L. Blair and many of his results are given in [We, Section 10]. Blair showed also that a countable union of z-embedded realcompact spaces is realcompact. Weir attributes the result cited above about Lindelöf spaces being z-embedded in any space containing it to M. Henriksen and D. G. Johnson who took care to attribute it to M. Jerison in [HeJo].

Such nice topological properties as normality and local compactness do not travel well between X and υX. Recall that a space X is called *pseudocompact* if $C(X) = C^*(X)$, or, equivalently, if $\upsilon X = \beta X$. The classical Tychonoff plank is pseudocompact and not normal, so the normality of X, and the existence of countably compact spaces that fail to be locally compact shows that local compactness need not descend from υX to X. Also A. Stone showed that X may be normal without υX being normal, and W. Comfort obtained a similar result for local compactness. See [We, Sections 8 and 10].

If this is not enough to shake your faith in paradise, read on.

2. The big problems; translation, representation, and characterization.

How does one take advantage of the fact that $C(X)$ determines X if X is realcompact? The first step is to be able to translate algebraic statements about $C(X)$ into topological statements about X and vice versa. Sometimes this is easy; e.g., it is an exercise to verify that f is invertible in $C(X)$ if and only if $Z(f)$ is empty, that f fails to be a proper divisor of zero if and only if Int $Z(f)$ is empty, and X is connected if and only if $C(X)$ has only the trivial idempotents $\mathbf{0}$ and $\mathbf{1}$. (Thus X is connected if and only if υX is connected).

Before giving some other translations that are less obvious, the order properties of $C(X)$ must be discussed. If we let

$$(f \wedge g)(x) = \min(f(x), g(x)), \text{and}$$

$$(f \vee g)(x) = \max(f(x), g(x))$$

for each $f, g \in C(X)$ and $x \in X$, then $C(X)$ is a lattice whose induced partial order is given by $f \geq g$ if $f(x) \geq g(x)$ for all $x \in X$. Clearly f and $g \geq 0$ imply $(f+g) \geq 0$ and $(fg) \geq 0$, so $C(X)$ is a lattice - ordered ring which is a subdirect product of copies of $\Re$. So $C(X)$ is an f-ring; see [BKW, Chap. 9]. Since $f \geq 0$ in $C(X)$ if and only if $f = g^2$, the order on $C(X)$ is an algebraic invariant, the lattice structure of $C(X)$ is determined by its structure as a ring. (A limited converse also holds; see below). A sample list of topological translations of algebraic properties follows.

$C(X)$	X
Every prime ideal of $C(X)$ is maximal	X is a P-space (i.e., every G_δ in X is open); see [GJ1,Chap4].
Every finitely generated ideal of $C(X)$ is principal	X is a F-space (i.e., every zero set of X is C^*-embedded); see [GJ1, Chap 14].
Every finitely generated ideal containing a nondivisor of 0 is principal	X is a quasi-F-space (i.e., every dense cozero set of X is C^*-embedded); see [DHH, Sec. 5].
$C(X)$ is a conditionally complete lattice	X is extremally disconnected [ED]; (i.e., the closure of every open set is open); see [GJ1, 3N]
$C(X)$ is conditionally σ-complete as a lattice	X is basically disconnected [BD]; (i.e., the closure of every cozero set is open); see [GJ1,3N].

The following diagram of implications holds

$$\begin{array}{l} P \\ \quad\searrow \\ ED \rightarrow BD \rightarrow F \rightarrow quasi-F \end{array}$$

and none of these implications hold in reverse; see [GJ1] and [DHH].

Notice that each of these "nice" algebraic properties translates into a topological property that most topologists consider pathological. This is typical, and the hiss of the serpent gets louder. As noted above, connectedness is equivalent to there being only trivial idempotents, and, within the class of realcompact spaces, compactness of X is equivalent to the existence for each f in $C(X)$ a positive integer n such that $|f| \leq n\cdot\mathbf{1}$. As will be noted below, some other "nice" topological properties have useful algebraic translations, but many others, like metrizability and local connectedness seem to lack nonponderous topological translations. It has been shown many times (for the latest such proof see [B]) that βX is locally connected if and only if X is locally connected and pseudocompact, (i.e., every continuous function is bounded). So any topological translation of local connectedness in the class of realcompact spaces must involve properties of $C(X)$ that do hold in $C^*(X)$; see [He 1].

Perhaps one of the more valuable uses of rings of continuous functions is as a way of representing archimedean lattice-ordered rings that are algebras over $\Re$ with an identity element e that is a weak order unit (i.e., $e \wedge x = 0$ implies $x = 0$). They are called Φ – *algebras* and can be represented as algebras of extended real-valued continuous functions as will be seen below.

If X is a compact space, let $\mathcal{D}(X)$ denote the family of continuous functions f on X into the two-point compactification $\gamma\Re = \Re \cup \{\pm\infty\}$ of $\Re$ such that $\{x \in X : f(x) \in \Re\}$ is a dense (open) subset of X. If $f \in \mathcal{D}(X)$, let $\mathcal{R}(f) = \{x \in X : f(x) \in \mathcal{R}\}$. If $f, g \in \mathcal{D}(X)$ and the sum of f and g on $\mathcal{R}(f) \cap \mathcal{R}(g)$ has a (necessarily unique) extension over X, then we say that $(f+g) \in \mathcal{D}(X)$. Similarly fg, $f \vee g$, and $f \wedge g$ are define when they are in $\mathcal{D}(X)$. While $\mathcal{D}(X)$ is always a lattice closed under scalar multiplication, it is usually not an algebra. Indeed $\mathcal{D}(X)$ is an algebra if and only if

X is a quasi-F-space; see [HeJo, 2.2]. The elements of $\mathcal{D}(X)$ are called *(continuous) extended real-valued functions* and it is shown that:

2.1 Theorem *Every Φ algebra* A *is isomorphic to an algebra $\bar{A}$ of extended real-valued functions on a compact space* X. *Moreover if K_1, K_2 are disjoint closed subsets of* X,*there is an $\bar{a} \in \bar{A}$ such that $\bar{a}[K_1] = \{0\}$ and $\bar{a}[K_2] = \{1\}$.*

Before outlining the proof of this representation theorem, we recall that an l-*ideal* of a lattice-ordered ring is the kernel of a homomorphism preserving both lattice and the ring operations. Every l-ideal I is a ring ideal such that if $b \in I$ and $|a| \leq |b|$, then $a \in I$. (In [GJ1], an l-ideal is called *absolutely convex*). Every l-ideal of a Φ-algebra is contained in a maximal l-ideal. If $\mathcal{M}(A)$ is the set of all maximal l-ideals endowed with the hull-kernel topology, then $\mathcal{M}(A)$ is compact. Moreover, for each $M \in \mathcal{M}(A), A/M = \Re$ or A/M contains $\Re$ properly, and we call M *real* or *hyper-real* accordingly. In either case, if $M \in \mathcal{M}(A)$ and $a \geq 0$, then $\bar{a}(M) = \inf \{r \in \Re : (a + M) \leq r\}$ is a real number or $+\infty$ (in case $(a+M) > r$ for each $r \in \Re$). Since A is archimedean, $\cap\{M \in \mathcal{M}(A) : \bar{a}(M) \in \Re\} = \{0\}$, so $\bar{a} : \mathcal{M} \to \Re \cup \{+\infty\}$ is real-valued on a dense subset of $\mathcal{M}(A)$, and turns out to be continuous as well. If $a \in A$ is arbitrary and $M \in \mathcal{M}(A)$, let $\bar{a}(M) = \bar{a}^+(M) - \bar{a}^-(M)$, where $a^+ = a \vee 0$ and $a^- = (-a) \vee 0$. Since $a^+ \wedge a^- = 0, \bar{a}$ is a continuous extended real-valued function on $\mathcal{M}(A)$ and the map $a \to \bar{a}$ is the desired isomorphism.

This result was improved upon by D. Johnson and J. Kist, and both results may be derived from the Yosida representation Theorem; see [LZ; Chapter 4].

This representation theorem has been used to characterize $C(X)$ algebraically within the class of Φ-algebras in an internal way. A Φ-algebra A is said to be *uniformly closed* if its image in $\mathcal{D}(\mathcal{M}(A))$ is closed under uniform convergence. (This is an algebraic concept since closure under uniform convergence is equivalent to saying that every Cauchy sequence of elements of A converges). It is noted in [HeJo] that A is uniformly closed if and only if $\bar{A}$ is an order-convex subalgebra of $\mathcal{D}(\mathcal{M}(A))$ containing $C(\mathcal{M}(A))$. The Φ-algebra A is called a *Φ-algebra of real-valued functions* if $\cap\mathcal{R}(A) = \{0\}$, where $\mathcal{R}(A)$ denotes the set of real maximal l-ideals of A. If $a \in A$, let $Z(a) = \{M \in \mathcal{M}(A) : |a| + M \leq \frac{1}{n}e + M \text{ for } n = 1, 2, \ldots\}$. A is said to be *closed under inversion* if for each $a \in A$ the smallest l-ideal containing a is all of A whenever $Z(a) \cap \mathcal{R}(A) = \emptyset$. The following lemma is usually at the crux of most characterization theorems; see [HeJo, 5.2].

2.2 Lemma A *Φ-algebra A is isomorphic to $C(Y)$ for some Tychonoff space Y if and only if*

(i) *A is an algebra of real-valued functions,*

(ii) *A is uniformly closed,*

(iii) *A is closed under inversion, and*

(iv) *$\Re(A)$ is z-embedded in $\mathcal{M}(A)$.*

While conditions (i), (ii), and (iii) are algebraic, and insure that A is isomorphic to an order - convex subalgebra of $C(\mathcal{R}(A))$, condition (iv) is doubly unsatisfactory from our point of view. It not only fails to be algebraic in character, but it refers to properties of the external object $C(\mathcal{R}(A))$.

This lemma or variations on it may be used easily to characterize $C(X)$ in case X is compact (replace (iv) by requiring that if $a \in A$, then $|a| \leq n \cdot e$ for some positive integer n), and in case X is locally compact and σ-compact (replace (iv) by requiring that $\mathcal{R}(A) = \mathcal{R}(h)$ for some $h \in A$). Characterizations for other classes of (realcompact) Tychonoff spaces are mentioned in [HeJo]. There is a characterization of $C(X)$ for X for any realcompact X due to F. Anderson [Ande] and corrected by G. Jensen in [Jen], but no algebraist I know can imagine verifying its complicated hypothesis in any concrete instance. Many other characterizations have been given, too numerous to mention in toto. Having many more beauties to choose from than did Paris, I single out a few as particularly worthy.

The first is due to D. Plank. He calls a Φ-algebra *normal* if each of its l-ideals that is closed under uniform convergence is contained in a real maximal l-ideal. In [Pl], he establishes the following lovely result.

2.3 Theorem (Plank) *A Φ-algebra A is isomorphic to $C(Y)$ for some Lindelöf space Y if and only if A is a uniformly closed Φ-algebra of real-valued functions that is closed under inversion and normal.*

Part of the reason that Plank's Theorem holds is that fact that a Lindelöf space is z-embedded in any Tychonoff space containing it (See Section 1).

To see how for condition (i), (ii), and (iii) of Lemma 2.2 fall short of characterizing a $C(X)$, in the general case, observed as in [HeJo] that the Φ-algebra B of Baire functions on $\Re$ into $\Re$ satisfies (i), (ii), and (iii) of Lemma 2.2, while $\mathcal{R}(B)$ is the set $\Re$ with the discrete topology. Thus, while B has power c, the algebra $C(\mathcal{R}(B))$ has power 2^c.

In 1975, A. Hager attempted to characterize $C(Y)$ for an arbitrary realcompact space Y in a succinct way in which will be described next. It remains doubtful that he succeeded, but his attempt has great value.

An ideal D of a Φ-algebra A is called *dense* if its only annihilator is 0. For any subset S of A, let coz $S = \cup\{\text{coz } \bar{a} : \bar{a} \in S\}$, where $a \rightarrow \bar{a}$ is the map of A into $\mathcal{D}(\mathcal{M}(A))$ defined above. Note the ideal D of A is dense if and only if $\mathcal{R}(A) \subset \text{coz} D$.

A dense ideal is said to be $\Re$-*dense* if it is contained in no real maximal ideal. If, for each $\Re$-dense ideal D of A, every module homomorphism $\varphi \in \mathrm{Hom}_A(D, A)$ is given by multiplication by some fixed element of D, the Φ-algebra A of real-valued functions is said to be closed under *strong inversion.* A may be regarded as a subalgebra of $C(\mathcal{R}(A))$, and A is closed under strong inversion if and only if whenever for $a, b \in A$, the quotient of a/b is defined on a neighborhood of each $x \in \mathcal{R}(A)$, we have $a/b \in A$. Modifying Hager's notation slightly, we call a Φ-algebra that satisfies (i) and (ii) of Lemma 2.2 and is closed under strong inversion an SI - *algebra.* A. Hager made the following conjecture as long ago as 1967. In [Ha 2] he calls it a "working conjecture" and expresses doubt about its validity.

2.4 Conjecture (Hager) *A Φ-algebra A is isomorphic to $C(X)$ for some realcompact space X if and only if A is an SI-algebra*

By way of evidence in its favor, Hager showed that every $C(X)$ is an SI-algebra. Moreover if $A = C(X)$ for some realcompact X, then we may assume $X = \mathcal{R}(A)$. A space X is said to be G_δ-*closed* in a space Y if for each $y \in Y$, there is a G_δ - set (or, equivalently a zeroset) of Y containing y and missing X. Note that X is real compact if and only if it is G_δ-closed in βX. To facilitate the rest of the discussion about Hager's conjecture, I introduce the following definition.

2.5 Definition A compactification K of a space X is called *large* if $K = \beta V$ whenever V is an open neighborhood of X in K.

Clearly βX is a large compactification of X and is clearly the only large compactification of X if X is locally compact or Lindelöf. Consider the following statements

(*) *If K is a large compactification of X, then $K = \beta X$, and*

(**) *If K is a large compactification of X and if X is G_δ-closed in K, then $K = \beta X$*

According to M. A. Sola, Hager conjectured that (*) holds as early as 1970, and in [Hal, 7,2], Hager shows that 2.4 holds if and only if (**) holds. It is doubtful that Hager conjectured (*), but it is still an interesting question.

In his University of South Carolina doctoral dissertation [S, Chap. 5] Sola shows that (*) need not hold with an example that does not disprove (**), so Hager's conjecture in 2.4 remains open.

What Sola does is to show that Prabir Roy's example Δ of a zero-dimensional metrizable space of power c for which $\beta\Delta$ fails to be zero dimensional has as a large compactification the maximal zero dimensional compactification $\zeta\Delta \neq \beta\Delta$. (The space ζX may be regarded as a space of ultrafilters on the collection of clopen subsets of Δ; see, for example [PW, Section 4.7], where $\zeta\Delta$ is denoted by $\beta_o\Delta$). Roy's example was announced first in [Roy] in 1962. It is complicated in nature; for a relatively easy development of its properties, see [Pe, Sect 7.4] (where Δ is denoted by P).

As Sola also observes, Δ is not G_δ-closed in the large compactification $\zeta\Delta$. For, as was shown by P. Nykos in 1971, (see [Ny]), Δ fails to be N-compact (i.e., Δ cannot be embedded as a closed subspace of a product of copies of the countable discrete space N) and it follows that there is a space $N\Delta$ properly in between Δ and $\zeta\Delta$ such that every G_δ of $\zeta\Delta$ containing a point of $N\Delta\backslash\Delta$ meets Δ. ($N\Delta$ is the N-compactification of Δ; see [Ny] or [S, Chapter 4]).

Thus Hager's conjecture remains unresolved. Most people who have worked on the problem (including Hager) guess that it is false. If so, a counterexample will be hard to find. The space Δ, being metrizable of nonmeasurable cardinal is realcompact - as will be any counterexample to (**).

Is the class of spaces for which (**) holds productive and/or closed hereditary- as is the class of realcompact spaces? *Can (**) be used as a unifying principle from which characterizations of $C(X)$ for several classes of realcompact spaces can be derived fairly easily?*

Surely, Hager's work in [Ha 1] should be exploited further.

In the same spirit, I mention G. DeMarco's characterization of $C(X)$ for X paracompact or strongly paracompact [DeM]. Recall that X is *paracompact* (resp. *strongly paracompact*) if every open cover of X has an open locally finite (resp. star finite) open refinement. A locally Lindelöf space is strongly paracompact if and only if it is a topological sum of Lindelöf spaces, but there are strongly paracompact spaces that fail to be locally Lindelöf. De Marco shows that a Φ-algebra A is isomorphic to a $C(X)$ for some strongly paracompact realcompact space if and only if A is an SI-algebra and each $\Re$-dense ideal of A contains an $\Re$-dense ideal that is a projective A-module. The reader is referred to [DeM] for the technically more complicated characterization in the paracompact case.

In a study related to the problem of characterizing $C(X)$, A. Hager and D. Johnson call a uniformly closed subalgebra of $C(X)$ that separates points from disjoint closed sets, contains all constant functions, and is closed under inversion an ALGEBRA *on* X. (The capitalization is used here to avoid confusion with the terminology used above; Hager and Johnson stick to lower case). They show in [Ha Jo] that if vX is Lindelöf, then any ALGEBRA on X is a $C(Y)$ for some Y and they asked if the converse holds. In [Bl 1] and [Bl 2], J. Blasco shows that it does if either X is a topological group or X is paracompact. In the general case, their question remains open for a more detailed study of the structure of ALGEBRAS, see [Ha 2].

I close this section with a mention of another use of algebraic properties of $C(X)$

to obtain information about X. A space X is called *rimcompact* if each point of X has a base of neighborhoods with compact boundaries. Each rimcompact space has a compactification ΦX such that

(a) $\Phi X \backslash X$ is zero-dimensional, and

(b) if ΨX is any compactification of X such that $\Psi X \backslash X$ is zero-dimensional, then there is a continuous map of ΦX onto ΨX that extends the identity map of X onto X.

The space ΦX is called the *Freudenthal compactification* of X and is essentially unique.

Let $C^{\#}(X) = \{f \in C(X)$: for each $M \in \mathcal{M}(C(X))$, there is an $r \in \Re$ with $(f - \mathbf{r}) \in M$. The subalgebra $C^{\#}(X)$ is a subalgebra of $C^{*}(X)$ and consists only of closed mappings of X into $\Re$ if X is rimcompact and realcompact. In different notation it was introduced by N. Shilkret in [Sh] who called it the *Gelfand subalgebra*, and independently L. Nel and Riordan in [NR]. In [He 2], I showed that if $C^{\#}(X)$ is rich enough to separate points from disjoint closed sets, i.e., to determine a compactification of X, and X is realcompact, then X is rimcompact and the compactification is ΦX, and I include an example due to S. Willard to show that $C^{\#}(X)$ need not determine a compactification of X even if X is both rimcompact and realcompact. The question of exactly when $C^{\#}(X)$ determines ΦX remains open. One may also ask if some related subalgebra of $C(X)$ may be found which determines ΦX whenever X is both rimcompact and realcompact.

There is a surprisingly large literature on this seemingly arcane area. It is summarized and improved upon by R. André in his University of Manitoba master's Thesis [Andr]. See also the Dickman-McCoy monograph on the Freudenthal compactification and the references therein [DM]. Also, in [Dom], J. Dominquez studies the Gelfand subalgebra for continuous functions with values in a nonarchimedean field.

3. H-fields and residue class fields of prime ideals; the really tough problems

The structure of H-fields has been studied a lot in the last 40 years. Much has been learned, but there are still many open problems. The latter seem to be quite difficult and seem to require heavy use of set theory, model theory and study of the structure of ultrafilters for their solution.

In [GJ1, Chap 13], it is shown that for every maximal ideal M of $C(X)$, the residue class field $C(X)/M$ is real-closed. Since much of what we know about an H-field depends heavily on its structure as an ordered set, I pause to review some definitions.

A cardinal number $\aleph_\alpha$ is *regular* if ω_α contains no cofinal subset of smaller cardinality; otherwise it is *singular*. If $\aleph_\alpha$ is regular, then $\alpha = \beta + 1$ for some cardinal β. More generally, the least cardinal number of a cofinal subset of a cardinal β is denoted by $cf(\beta)$. Thus, β is regular if and only if $cf(\beta) = \beta$. If A and B are subsets of a (totally) ordered set L, and $a \in A$ and $b \in B$ imply $a < b$, we write $A < B$.

3.1 Definition *Suppose L is an ordered set and α is an ordinal number. If, whenever A and B are subsets of L of cardinality less than $\aleph_\alpha$ such that $A < B$, there is an x in L such that $A < x < B$, the set L is called an η_α - set.*

Since A or B may be empty, an η_α-set has no cofinal or coinitial subset of power smaller than $\aleph_\alpha$. The ordered set Q of rational numbers is the only η_o-set up to an order isomorphism; and it is known that any ordered set of power $\aleph_\alpha$ may be embedded in an η_α-set; see [GJ1 13O].

This section depends heavily on the theory of ultrafilters and [CN] is a principal reference, so it must be pointed out that my terminology conflicts with theirs; what I call an η_α-set is called an η_{ω_α}-set in [CN].

In [EGH] and [GJ1, Chap. 13], it is shown that every H-field is an η_1-set. An ordered field that is an η_α-set is called an *η_α-field.* In this language, every H-field is a real-closed η_1-field. It was shown also in [EGH] that *any two real-closed η_α-fields of cardinality $\aleph_\alpha$ are isomorphic if $\alpha > 0$.* It is noted that there are no η_α-sets of power $\aleph_\alpha$ if $\aleph_\alpha$ is singular, and if there is an $\eta_{\alpha+1}$ - set of cardinality $\aleph_{\alpha+1}$, then $2^{\aleph_\alpha} = \aleph_{\alpha+1}$. This implies that if CH holds, then two real-closed fields of power c are isomorphic. Jumping a bit ahead in our story, A. Dow in [Dow] (improving on consistency results in [ACCH] and [Ro]) showed that the converse holds as well; if CH fails, then there is a pair of nonisomorphic H-fields of power c.

An H-field that is a homomorphic image of $C(D)$ where D is a discrete space is called an *ultrapower* of $\Re$. In [EGH] it is shown that there are ultrapowers of arbitrarily large cardinality. Indeed, all of the H-fields exhibited in [EGH] are ultrapowers, as are the H-fields exhibited by A. Dow in [Dow, 2.3] in the absence of CH. Dow's H-fields are homomorphic images of $C(N)$.

The results in [EGH] led to a variety of questions only some of which were posed explicitly in that paper.

If it were true that every H-field is an η_α-field of power $2^{\aleph_\alpha}$ for some $\alpha > 0$, then each H-field would be determined by its cardinal number at least if GCH (the generalized continuum hypothesis) holds. It is asked in [EGH, 5.2] if this latter is true. Note that its truth would imply that there is no H-field of singular cardinality.

It is not obvious that there are any real-closed η_α-fields for any $\alpha > 1$, and this is listed as an open problem in [GJ1, 13O]. In a paper that has largely been forgotten by workers on H-fields, N. Alling constructed in his 1962 paper [Al 1] for each regular cardinal $\aleph_{\alpha+1}$ a real-closed $\eta_{\alpha+1}$-field of power $\aleph_{\alpha+1}$ under the necessary assumption that $2^{\aleph_\alpha} = \aleph_{\alpha+1}$. In another not well-known expository paper present in Rome, Italy in 1973, [Al 2] Alling summarized the state of knowledge about H-fields at that time. He reported that Keisler had shown if GCH hold, and $\alpha > 0$, then every real-closed $\eta_{\alpha+1}$ field of power $\aleph_{\alpha+1}$ is an ultrapower, but not every ultrapower takes this form. Keisler's results in [Ke] were phrased in the language of model theory and hence had not been read by many workers in $C(X)$. In particular, η_α-fields are called $\aleph_\alpha$-*saturated fields*, so the applicable results even in [CN, pp. 320-325] are hard to recognize.

A massive attack by four very capable mathematicians on the problems of determining the structure of H-fields was made in the mid-1970's. While some of their results were presented at the Prague topology conference in 1976, [ACCH] did not appear until 1981. Below, I list some of its major results.

3.2 *m is the cardinal number of an H-field if and only if $m = m^{\aleph_0}$.*

If a fragment of GCH holds, then it follows from [Jec, Corollary 2, p. 49] that $\aleph_{\omega_1}$ is a singular cardinal satisfying 3.2. As pointed out to me by W. Comfort, to get an example in ZFC, it is enough to take a beth cardinal that is a strong limit cardinal of uncountable cofinality; see [ACCH, Sect 1] or [CN, Chap. 1] for the appropriate definitions.

In [ACCH, 7.7], it is shown that every H-field is a homomorphic image of $C(\Re^\Delta)$ for some cardinal Δ, and, assuming GCH, a cardinal number is exhibited that satisfies the conditions of 3.2 above that fails to be the cardinal number of any ultrapower;

loc. cit., 5.11.

If M is a hyper real ideal of $C(X)$, let m denote the least cardinal number of a zero set in $\mathcal{Z}(M)$. Must the cardinality of C/M exceed m? Having answered this in the affirmative if $m = \aleph_o$ or c, or if X is discrete, the problem is posed in 5.3 of [EGH], and is solved in the negative in ZFC in [ACCH, Section 6], where it is pointed out that 5.3 had been solved earlier by a variety of authors under set theoretic assumptions.

In 5.4 of [EGH] the reader is asked to investigate to what extent results on H-fields depend on CH. Dow answers one such question definitively in [Dow] as noted above, and in [ACCH] many more problems of this sort are considered.

[ACCH] is full of other interesting results which the reader is urged to examine in the original. These authors also pose many intriguing questions. I repeat only three of them.

3.3 *Is every real-closed η_1-field an H-field?*

3.4 *Are order isomorphic H-fields algebraically isomorphic?*

3.5 *Is there any way to connect the cardinal number of the H-field $C(X)/M$ with topological or combinatorial properties of X and $\mathcal{Z}(M)$ respectively?*

I hope that this summary serves to generate more interest in the study of H-fields. Success in this area will require expertise in set theory, model theory, topology, and algebra. Perhaps a lot of team efforts will be needed.

To facilitate the ideas presented in the rest of this section, I introduce the following definition

3.6 Definition *If X is a topological space and P is a prime ideal of $C(X)$ such that $C(X)/P$ contains $\Re$ properly, then $C(X)/P$ is called an H- domain.*

While this terminology appears here for the first time, the study of H-domains is far from new. Every H-domain D is a totally ordered integral domain whose field of quotients is real-closed since every positive element has a square root and every monic polynomial of odd degree has a zero in D. (This was observed by N. Alling in 1963 in [Al3]). $C(X)/P$ is a H-domain as long as P is a nonmaximal prime ideal of $C(X)$, and the set of prime ideals of $C(X)/P$ is totally ordered; it is $\{Q/P : Q$ prime and $Q \supset P\}$. C. Kohls studied prime ideals of $C(X)$ and the order structure of H-domains in his Purdue doctoral dissertation which is most of the content of [GJ1, Chap 14]; see also [Ko 1, 2, 3, 4]. In these studies, as in [M 1,2], the emphasis is more on prime ideals than on the algebraic structure of H-domains. Until recently, I have seen very few studies of the structure of H-domains that are not fields. One worthy of mention is [GJ 2] where some necessary and some sufficient conditions are given in order that the quotient field of an H-domain be a H-field. It rests heavily on the contents of [GJ1, Chap. 14].

In the last few years two substantial studies of H-domains have been made. In [CD 1], a (commutative) integral domain A is called a *real-closed ring* if it is an ordered ring satisfying the intermediate value property for polynomials. That is, if $a, b \in A, a < b$, and $f(X) \in A[x]$ satisfies $f(a)f(b) < 0$, then there is a $c \in A$ in the open interval (a, b) such that $f(c) = 0$. A prime ideal P of $C(X)$ is called a *real-closed ideal* if $C(X)/P$ is a real-closed ring. A commutative totally ordered integral domain A is a real-closed ring if and only if $0 < a < b$ implies a divides b; that is if and only if A is a *valuation ring* (i.e., of any two elements, one must be divisible by the other). It turns out that if P is a real-closed ideal, and Q is a prime ideal containing P, then Q is a real-closed [CD 1, Prop. 7]. Since every minimal prime ideal of $C(X)$ is a z-ideal, it suffices to determine when z-ideals are real-closed. A technical topological characterization of real-closed z-ideals which proves to be a useful tool is given in [CD 1, Theorem 1 and Prop. 5]. A few of their results follow:

If X is an F-space, then every prime ideal is real-closed, but the converse fails. If X is metrizable, $p \in X$, and $M_p = \{f \in C(X) : f(p) = 0\}$ contains a nonmaximal real-closed ideal, then there is a P-point of $\beta N - N$, i.e., a point in the interior of any G_δ containing it. It is known that there are models of set theory in which $\beta N - N$ has no P-point and Martin's Axiom implies their existence; see [vM]. A detailed study of the relationship between the real-closed ideals of the ring $C(N^*)$ of convergent sequences of real numbers and the P-points of $\beta N - N$ is made, and the set of real-closed ideals of $C(D^*)$ (where D^* is the one-point compactification of an uncountable discrete space) and of $C[0,1]$ is studied carefully. Many unsolved problems remain. Many of the questions asked about H-fields may also be asked about H-domains that are real-closed rings. For example, does the order type of such a ring determine it to within an algebraic isomorphism?

[CD 2] is concerned mainly with model-theoretic questions about real-closed rings. In his Rutgers doctoral dissertation, J. Moloney showed that up to isomorphism, there are exactly 10 H-domains of $C(N^*)$. CH is assumed freely in his work, and to the best of my knowledge, nobody has examined whether these results hold with less restrictive set-theoretic assumptions. He also classifies some of the domains that are homomorphic images of $C^\infty([0,1])$. These results are available in preprint form [Mo].

4. Miscellania

This section is about a few topics near the boundary of the scope of this paper which merit at least brief mention.

I. Discontinuous homomorphisms of commutative Banach algebras

If X is a compact space, and we let $\| f \| = \sup \{| f(x) |: x \in X\}$ for each $f \in C(X)$, then $(C(X), \| \cdot \|)$ is a commutative semisimple Banach algebra. In 1949, I. Kaplansky asked if there could be a norm $\| \cdot \|'$ on a $C(X)$ with respect to which it becomes a normed algebra (satisfying $\| fg \|' \leq \| f \|' \| g \|'$), with respect to which $C(X)$ fails to be complete. This problem reduces to the questions of whether there is a discontinuous homomorphism of $(C(X), \| \cdot \|)$ into a Banach algebra. It was 25 years before this question could be answered for any infinite compact X, and then, independently, G. Dales and J. Esterle showed that if CH holds and X is a compact space then $C(X)$ has a discontinuous homomorphism. Shortly thereafter, R. Solovay and his student, H. Woodin, announced that there are models for set theory in which whenever X is compact, every homomorphism of $C(X)$ into a Banach algebra is continuous. In a recent book [DW], Dales and Woodin provide an excellent exposition of the Solovay-Woodin independence proofs making these results accessible for the first time to mathematicians that are not experts in model theory. If you want to learn how to use Martin's Axiom and forcing, this is a great place to start. It also contains a complete set of references and a history of the problem.

II. Other algebraic structures on $C(X)$

Probably many of you are more familiar with the structure of $C(X)$ as a Riesz space (alias vector-lattice) than as an algebra. Indeed, an archimedean Riesz space with a weak order e admits at most one multiplication making it into a Φ-algebra in which e is the identity element. More generally, it always has a unique embedding into a Φ-algebra. The nature of the embedding and the way in which a change of the weak order unit can effect the multiplicative structure is described in [HR 1,2]. This work overlaps with earlier work of P. Conrad on how the multiplication of an f-ring is determined by its structure as an abelian l-group (not necessarily uniquely in the nonarchimedean case); see [Co].

In 1956, making use of T. Shirota's generalization of earlier work of I. Kaplansky and A. Milgram, I thought I had shown that $C(X)$ either as a lattice or as a multiplicative semigroup determined υX; see [He 3]. In [Cs] A. Csazar points out that there is a gap in Shirota's reasoning and provides a much simpler way of showing that $C(X)$ as a multiplicative semigroup determines X. Indeed, he finds that υX is determined by semigroup structures that are much simpler.

Despite this, I find myself more comfortable working with the ring structure in preference to that of Riesz spaces, lattices, or multiplicative semigroups. That may, however, just be a personal prejudice.

III. Spaces of Minimal Prime Ideals

Since 1965, there has been a lot written on the space of minimal prime ideals of a commutative ring with the hull-kernel topology. In case the ring is a $C(X)$ the space mX of minimal prime ideals of $C(X)$ had been known to be countably compact since 1965 and was known to be basically disconnected if it is locally compact; see [He Je] and [Ki]. Recently, in two papers [DHKV1,2], it has been shown that mX is always ω-bounded, indeed that weakly Lindelöf subspaces of mX have compact closures and are C^*-embedded in mX, but mX need not be basically disconnected, or even a quasi-F-space. It is not known exactly when mX is basically disconnected, and it would be nice to get a topological characterization on mX.

IV. Covers and completions

In 1958, A. Gleason showed that the projective objects in the category of compact spaces and continuous maps are extremally disconnected, and there is for each X

in this category a compact extremally disconnected space minimal with respect to mapping continuously onto X. By now most authors call this space the *absolute* of X and denote it by EX. The ring $C(EX)$ turns out to be the Dedekind-MacNeille completion of $C(X)$.

Parts of this theory can be extended to the category of Tychonoff spaces and perfect maps (A map is *perfect* if it closed, continuous, and the inverse images of each point is compact), but in general the Dedekind completion of a $C(X)$ need not be a $C(Y)$; see [MJ].

Other kinds of completions of a $C(X)$ may be obtained in similar ways, but each different kind of completion has its own complications. Those who would like to learn more may examine [DHH], [Ha3], [HVW] and [Z]. Excellent expositions of the theory of absolutes are given in [Wa] and [PW].

The Henriksen-Johnson representation Theorem 2.1 above has been used to extend many results about $C(X)$ to archimedean Φ-algebras in appropriately modified form. This could be the subject of another talk at least as long.

I hope I have convinced you that while working in this area will not put you in a mathematical paradise, it can fulfill your needs for working on challenging and interesting problems.

References

[Al 1] N. Alling, *The existence of real-closed fields that are η_α-sets of power $\aleph_\alpha$*, Trans. Amer. Math Soc. 103 (1962), 341-352.

[Al 2] ___________, *Residue class fields of rings of continuous functions*, Symposia Mathematica 17 (1976), 55-67.

[Al 3] ___________, *An application of valuation theory to rings of continuous real and complex-valued continuous functions*, Trans. Amer. Math. Soc. 109 (1963), 492-508.

[Ande] F. Anderson, *Approximation in systems of real-valued functions*, Trans. Amer. Math. Soc. 103 (1962), 249-271.

[Andr] R. André, *On compactifications determined by $C^\#(X)$*, Master's Thesis, Univ. of Manitoba, 1987.

[ACCH] M. Antonovskij, D. Chudnovsky, G. Chudnovsky, and E. Hewitt, *Rings of real-valued continuous functions* II, Math. Zeit. 176 (1981), 151-186.

[AS] R. Alo and H. Shapiro, *Normal Topological Spaces*, Cambridge University Press, Great Britain, 1974.

[Bl 1] J. Blasco, *A note on inverse closed subalgebras*, Acta. Math. Hung. 51 (1988), 47-49.

[Bl 2] ___________, *Complete bases in topological spaces*, Studia Sci. Math. Math. Hung. 19 (1984), 49-54.

[Bo] D. Baboolal, *Local connectedness in the Stone-Čech compactification*, Canad. Math. Bull 31 (1988), 236-240.

[BKW] A. Bigard, K. Keimel, and S. Wolfenstein, *Groupes et Anneaux Réticulés*, Springer-Verlag Lecture Notes in Mathematics 609, New York 1977.

[Co] P. Conrad, *The additive group of an f-ring*, Canad. J. Math. 26 (1974), 1157-1168.

[Cs] A. Csazar, *Semigroups of continuous functions*, Acta. Sci. Math. Hung. 45 (1983), 131-140.

[CD 1] G. Cherlin and M. Dickmann, *Real closed rings* I, Fund. Math. 126 (1986), 147-183.

[CD 2] ____________, *Real-closed rings* II, Ann. Pure and Applied Logic 25 (1983), 213-231.

[CN] W. Comfort and S. Negrepontis, *The Theory of Ultrafilters*, Springer-Verlag, New York 1974.

[DeM] G. De Marco, *A Characterization of $C(X)$ for X strongly paracompact (or paracompact)*, Symposia Math. 21 (1977), 547-554.

[Dom] J. Dominquez, *Nonarchimedean $C^{\#}(X)$*, Proc. Amer. Math. Soc. 97 (1986) 525-530.

[Dow] A. Dow, *On ultra powers of Boolean algebras*, Top. Proc. 9 (1984), 269-291.

[DHH] F. Dashiel, A. Hager, and M. Henriksen, *Order - Cauchy completions of rings and vector lattices of continuous functions*, Canad. J. Math. 32 (1980) 657-685.

[DM] R. Dickman and R. McCoy, *The Freudenthal Compactification*, Math. Dissertations 162, Warsaw 1988.

[DHKV 1] A. Dow, M. Henriksen, R. Kopperman, and H. Vermeer, *The space of minimal prime ideals of a $C(X)$ need not be basically disconnected*, Proc. Amer. Math. Soc., to appear.

[DHKV 2] ____________, *The countable annihilator condition and weakly Lindelöf subspaces of minimal prime ideals* Proc. Baku Topology Conference, to appear.

[DW] H. Dales and H. Woodin, *An Introduction to Independence Proofs for Analysts*, London Math. Soc. Lecture Notes 115, Cambridge University Press 1987.

[EGH] P. Erdos, L. Gillman, and M. Henriksen, *An isomorphism theorem for real-closed fields*, Annals of Math. 6 (1955), 542-554.

[G] L. Gillman, *Rings of continuous functions are rings*, Rings of Continuous Functions, 143-147, Marcel Dekker Inc., New York 1985.

[GJ 1] L. Gillman and M Jerison, *Rings of Continuous Functions*, Van Nostrand Publ. Co., Princeton 1960.

[GJ 2] ____________, *Quotient fields of residue class rings of continuous functions*, Ill, J. Math. 4 (1960), 425-436.

[Ha 1] A. Hager, *A class of function algebras (and compactifications and uniform spaces)* Sumposia Mathematica 17 (1976), 11-22.

[Ha 2] ____________, *On inverse-closed subalgebras of* $C(X)$, Proc. London Math Soc. 19 (1969), 233-257.

[Ha 3] ____________, *Minimal Covers of topological spaces*, Proc. N.Y. Acad. Sci., to appear.

[He 1] M. Henriksen, *Unsolved problems on algebraic aspects of* $C(X)$, Rings of Continuous Funmctions, 195-202, Marcel Dekker Inc, New York 1985.

[He 2] ____________, *An algebraic characterization of the Freudenthal Compactification*, Top. Proc. 2 (1977), 169-178.

[He 3] ____________, *On the equivalence of the ring, lattice, and semigroup of continuous functions*, Proc. Amer. Math. Soc. 7 (1956) 959-960.

[Hew] E. Hewitt, *Rings of real-valued continuous functions* I, Trans. Amer. Math. Soc. 64 (1948), 45-99.

[Ho] R. Hodel, *Cardinal functions* I, Handbook of Set-Theoretic Topology, 3-61, North Holland, Amsterdam 1984.

[Ha Jo] A. Hager and D. Johnson, *A note on certain subalgebras of* $C(X)$, Canad. J. Math. 20 (1968), 389-393.

[He Je] M. Henriksen and M. Jerison, *The space of minimal prime ideals of a commutative ring*, Trans. Ameri. Math. Soc. 115 (1965), 110-130.

[He Jo] M. Henriksen and D. Johnson, *On the structure of a class of Archimedean lattice-ordered algebras*, Fund. Math. 50 (1961), 73-94.

[HR 1] A. Hager and L. Robertson, *Representing and ringifying a Riesz space* Symposia math. 21 (1977), 411-430.

[HR 2] ___________, *On the embedding into a ring of an archimedean l-group.* Canad. J. Math. 31 (1979), 1-8.

[HVW] M. Henriksen, J. Vermeer, and R.G. Woods, *Quasi- F-covers of Tychonoff spaces*, Trans. Amer. Math. Soc 303 (1987), 779-803.

[Jec] T. Jech, *Set Theory*, Academic Press New York 1978.

[Jen] G. Jensen, *A note on complete separation in the Stone Topology*, Proc. Amer. Math. Soc. 21 (1969) 113-116.

[Ke] H. Keisler, *Good ideals in fields of sets*, Annals of Math. 79 (1964), 338-359.

[Ki] J. Kist, *Minimal prime ideals in commutative semigroups*, Proc. London Math. Soc. 13 (1963), 31-50.

[Ko 1] C. Kohls, *Ideals in rings of continuous functions*, Fund. Math. 45 (1957) 28-50.

[Ko 2] ___________, *Prime ideals in rings of Continuous functions* Ill. J. Math 2 (1958) 505-536.

[Ko 3] ___________, *Prime ideals in rings of continuous functions* II, Duke Math. J. 25 (1958), 447-458.

[Ko 4] ___________, *Prime z-filters on completely regular spaces*, Trans. Amer. Math. Soc. 120 (1965), 299-306.

[Ma 1] M. Mandelker, *Prime ideal structure of rings of bounded continuous functions*, Proc. Amer. Math. Soc. 19 (1968), 1432-1438.

[Ma 2] ___________, *Prime z-ideal structure of* $C(\Re)$, Fund. Math. 63, (1968), 145-166.

[vM] J. van Mill, *An introduction to* $\beta\omega$, Handbook of Set-theoretic Topology, 503-567, North Holland 1984.

[MJ] J. Mack and D. Johnson, *The Dedekind completion of* $C(X)$, Pac. J. Math. 20 (1967), 231-243.

[N] P. Nykos, *Prabir Roy's space Δ is not N-compact*, Gen. Top and Appl. 3 (1973), 197-210.

[NR] L. Nel and D. Riordan, *Note on a subalgebra of $C(X)$*, Canad. Math. Bull 15 (1972), 607-608.

[Pe] A. Pears, *Dimension Theory*, Cambridge Univ. Press 1975.

[Pl] D. Plank *Closed l-ideals in a class of lattice-ordered algebras*, Ill. J. Math. 15 (1971), 515-524.

[PW] J. Porter and R. G. Woods, *Extensions and Absolutes of Hausdorff Spaces*, Springer-Verlag, New York 1987.

[Roi] J. Roitman, *Non-isomorphic hyper-real fields from non-isomorphic ultrapowers*, Math. Zeit. 181 (1982), 93-96.

[Roy] P. Roy, *Failures of equivalence of dimension concepts for metric spaces*, Bull Amer. Math. Soc. 68 (1962), 609-613.

[Sh] N. Shilkret, *Non-Archimedean Banach Algebras*, Duke Math. J. 37 (1970), 315-322.

[So] M. Sola, *Roy's Space Δ and ITS N-Compactification*, Doctoral Dissertation, Univ. of South Carolina, 1987.

[Wa] R. Walker, *The Stone-Čech compactification*, Springer-Verlag, New York, 1974.

[We] M. Weir, *Hewitt-Nachbin Spaces*, North Holland, 1975.

[Wo] R. Woods, *A Survey of Absolutes of Topological Spaces*, Topological Structure II, Math Centre Tracts 116 (1979), Amsterdam.

[Z] V. Zaharov, *On functions connected with sequential absolutes, Cantor completion and classical rings of quotients*, Periodica Math. Hung. 19 (1988), 113-133.

ON SOME CLASSES OF LATTICE-ORDERED ALGEBRAS

S.J. Bernau
Department of Mathematical Sciences
University of Texas
at El Paso
El Paso, Texas 79968-0514
U.S.A.

C.B. Huijsmans
Department of Mathematics
Rijksuniversiteit Leiden
P.O. Box 9512
2300 RA Leiden
The Netherlands

ABSTRACT. In this paper we present a survey of properties of Archimedean d-algebras and almost f-algebras. The main result is that every Archimedean almost f-algebra is commutative. Furthermore, a description is given of the nilpotents in both an Archimedean d-algebra and an Archimedean almost f-algebra. In both cases every nilpotent a satisfies $a^3 = 0$ and $bac = 0$ for all b, c.

The (associative) lattice-ordered algebra (ℓ-algebra) A is called an f-algebra (f for function) whenever $a, b \in A, a \perp b$ (i.e., $|a| \wedge |b| = 0$) implies $ca \perp b$ and $ac \perp b$ for all $c \in A$.
It is well-known by now that an Archimedean f-algebra is automatically both associative and commutative. For the various proofs of this result we refer to Amemiya [1, 1953], Birkhoff and Pierce [7, 1956], Bernau [3, 1965], Zaanen [13, 1975] and Huijsmans and de Pagter [8, 1986].
For any ℓ-algebra A, denote by $N(A)$ the set of all nilpotent elements in A; if $N(A) = \{0\}$, then A is called semiprime.
If A is an Archimedean f-algebra, then $N(A)$ is a band in A and satisfies

$$\begin{aligned} N(A) &= \{a \in A : a^2 = 0\} \\ &= \{a \in A : ab = 0 \text{ for all } b \in A\} \end{aligned}$$

(see e.g. [10, proposition 10.2] or [14, theorem 142.5]).

It is the main purpose of the present note to investigate how far the above f-algebra results hold true for more general classes of ℓ-algebras. To be more precise,

J. Martinez (ed.), Ordered Algebraic Structures, 175–179.

we shall consider the classes of almost f–algebras (as introduced by Birkhoff in [6]) and d–algebras (which notion seems to go back to Kudláček [9]).
This note is meant to be a survey of our results. All theorems are therefore stated without proofs. For a more extensive version we refer to a forthcoming paper [4].

Definition 1. (i) The ℓ–algebra A is said to be an almost f–algebra whenever $a, b \in A, a \perp b$ implies $ab = 0$.
(ii) the ℓ–algebra A is called a d–algebra whenever $a, b \in A, a \perp b$ implies $ca \perp cb$ and $ac \perp bc$ for all $c \in A$.

Almost f–algebras are called ff–algebras by Scheffold in [12]. An ℓ–algebra A is an almost f–algebra if and only if $a^2 = |a|^2$ for all $a \in A$. Hence, squares are positive in any almost f–algebra. The ℓ–algebra A is a d–algebra if and only if $|ab| = |a| \cdot |b|$ for all $a, b \in A$. Equivalently, $(ab)^+ = a^+b^+ + a^-b^-$ for all $a, b \in A$.
For the proof of the following theorem we refer to [7, section 9] and [5, section 9.1].

Theorem 2. (i) Any semiprime d–algebra is an f–algebra
(ii) any semiprime almost f–algebra is an f–algebra
(iii) any d–algebra with unit element $e > 0$ is an f–algebra
(iv) any Archimedean almost f–algebra with unit element e is an f–algebra.

Example 16 of [7] demonstrates that theorem 2 (iv) fails to hold if A is not Archimedean.

It was proved by Scheffold in [12] that every almost Banach f–algebra is automatically commutative. With this as a starting point, Basly and Triki showed in [2] that actually every Archimedean almost f–algebra is commutative. The disadvantage of both proofs however is that they rely heavily on representation theory, so Zorn's lemma (i.e., the axiom of choice) is (unneccesarily) involved. Recently, we succeeded to prove this result in an elementary representation-free way. For details, see [4].

Theorem 3. Every Archimedean almost f–algebra is commutative.

Rather surprisingly and contrary to the f–algebra case, an Archimedean almost f–algebra A is not automatically associative. By way of example, take $A = C\ ([0,1])$ with the pointwise vector space operations, partial ordering and the following (commutative) multiplication: if $a, b \in A$, define

$$(a * b)(x) = a(0)b(0) + a(1)b(1)$$

for all $x \in [0,1]$. Then A satisfies all the properties of an Archimedean almost f-algebra, except for the associativity.

An Archimedean d-algebra A need not be commutative. For instance, if $A = \mathbf{R}^2$, the coordinatewise ordered plane with the following multiplication:

$$a \cdot b = \begin{pmatrix} \alpha_1 \\ \alpha_2 \end{pmatrix} \cdot \begin{pmatrix} \beta_1 \\ \beta_2 \end{pmatrix} = \begin{pmatrix} \alpha_1 \beta_1 \\ \alpha_1 \beta_2 \end{pmatrix},$$

then A is an Archimedean d-algebra.

Observe in this connection that every commutative d-algebra is an almost f-algebra ([7, lemma 9.3]).

Now, we turn to the description of $N(A)$ in both an almost f-algebra and a d-algebra.
It was shown by Scheffold in [12] that

$$N(A) = \{a \in A : a^3 = 0\}$$

in any almost Banach f-algebra A. Once more by representation, Basly and Triki show in [2] that the latter formula holds true in arbitrary Archimedean almost f-algebras. Again, we were able to produce an elementary, representation-free proof of this equality and even more (see [4]).

Theorem 4. If A is an Archimedean almost f-algebra, then

$$\begin{aligned} N(A) &= \{a \in A : a^3 = 0\} \\ &= \{a \in A : a^2 b = 0 \text{ for all } b \in A\} \\ &= \{a \in A : abc = 0 \text{ for all } b, c \in A\}. \end{aligned}$$

The cube in the description of $N(A)$ is sharp, that is to say, in an almost f-algebra A which is not an f-algebra one cannot do with the square. In illustration, if we define in the coordinatewise ordered plane $A = \mathbf{R}^2$ the multiplication

$$a \cdot b = \begin{pmatrix} \alpha_1 \\ \alpha_2 \end{pmatrix} \cdot \begin{pmatrix} \beta_1 \\ \beta_2 \end{pmatrix} = \begin{pmatrix} 0 \\ \alpha_1 \beta_1 \end{pmatrix},$$

then A is an Archimedean almost f–algebra (and a d–algebra), but not an f–algebra. It is easily verified that $N(A) = A$, but $\{a \in A : a^2 = 0\}$ is the vertical axis.

Due to the fact that the multiplications in an f–algebra A are order continuous ([14, chapter 20]), $N(A)$ is a band in A. Though $N(A)$ is an ℓ–ideal in any Archimedean almost f–algebra A, it is not necessarily a band. If e.g. $A = C([0,1])$ with the pointwise vector space operations and partial ordering and the multiplication in A is for all $a, b \in A$ defined by

$$(a * b)(x) = a(0)b(0) \quad (0 \leq x \leq 1),$$

then A is an Archimedean almost f–algebra and a d–algebra, but not an f–algebra. Furthermore,

$$N(A) = \{a \in A : a(0) = 0\}$$

is not a band in A.

Concerning the description of $N(A)$ in Archimedean d–algebras, we note first that Quint showed in her thesis [11] that

$$\begin{aligned} N(A) &= \{a \in A : a^3 = 0\} \\ &= \{a \in A : cab = 0 \text{ for all } b, c \in A\} \end{aligned}$$

in any Banach d–algebra A. These formulas remain valid for arbitrary Archimedean d–algebras (commutative or not). For a proof, we refer again to [4].

Theorem 5. If A is an Archimedean d–algebra, then

$$\begin{aligned} N(A) &= \{a \in A : a^3 = 0\} \\ &= \{a \in A : a^2 b = 0 \text{ for all } b \in A\} \\ &= \{a \in A : ca^2 = 0 \text{ for all } c \in A\} \\ &= \{a \in A : cab = 0 \text{ for all } b, c \in A\}. \end{aligned}$$

This theorem has an interesting consequence. If A is an Archimedean non-commutative d–algebra, then the quotient $A/N(A)$ is an Archimedean semiprime d–algebra and hence an Archimedean f–algebra. It follows that $A/N(A)$ is commutative, so $ab - ba \in N(A)$ for all $a, b \in A$. This shows that

$$c(ab - ba)d = 0$$

for all $a, b, c, d \in A$. In particular (take $c = d = ab - ba$), $(ab - ba)^3 = 0$ for all $a, b \in A$. Also, $(ab)^2 = a^2 b^2$ for all $a, b \in A$ (take $c = a, d = b$).

REFERENCES

[1] I. AMEMIYA, A general spectral theory in semi-ordered linear spaces, J. Fac. Sc. Hokkaido Un. Ser. I, 12 (1953), 111-156.

[2] M. BASLY AND A. TRIKI, FF–Algèbres Archimédiennes réticulées, preprint, University of Tunis (1988).

[3] S.J. BERNAU, On semi-normal lattice rings, Proc. Camb. Phil. Soc. 61 (1965), 613-616.

[4] S.J. BERNAU AND C.B. HUIJSMANS, Almost f–algebras and d–algebras, to appear in Math. Proc. Camb. Phil. Soc.

[5] A. BIGARD, K. KEIMEL AND S. WOLFENSTEIN, Groupes et anneaux réticulés, Lecture Notes 608 (Springer, 1977).

[6] G. BIRKHOFF, Lattice theory, A.M.S. Coll. Publ. 25, 3rd edition (1967).

[7] G. BIRKHOFF AND R.S. PIERCE, Lattice-ordered rings, An. Acad. Bras. Ci. 28 (1956), 41-69.

[8] C.B. HUIJSMANS AND B. DE PAGTER, Averaging operators and positive contractive projections, J. Math. Anal. Appl. 113 (1986), 163-184.

[9] V. KUDLÁČEK, On some types of ℓ–rings, Sborní Vysokého Učeni Techn. v Brně (1962), 179-181.

[10] B. DE PAGTER, F–Algebras and orthomorphisms, Ph.D. Thesis, Leiden (1981).

[11] C. QUINT, Zur Darstelling von Banachverbandsalgebren, Ph.D. Thesis, Darmstadt (1984).

[12] E. SCHEFFOLD, FF–Banachverbandsalgebren, Math. Z. (1981), 193-205.

[13] A.C. ZAANEN, Example of orthomorphisms, J. Appr. Theory 13 (1975), 192-204

[14] A.C. ZAANEN, Riesz spaces II (North-Holland, Elsevier, 1983).

Primary ℓ-ideals in a class of f-rings

Suzanne Larson
Loyola Marymount University

An f-ring is a lattice ordered ring which is a subdirect product of totally ordered rings. General information on f-rings can be found in [**1**]. We will assume throughout that all of our f-rings are commutative and semiprime. Let A be an f-ring and I, J be ℓ-ideals of A. Recall that I is *semiprime (prime)* if $a^2 \in I$ ($ab \in I$) implies $a \in I$ ($a \in I$ or $b \in I$). Let $\sqrt{I}$ denote $\{a \in A : a^n \in I \text{ for some } n\}$, the smallest semiprime ℓ-ideal containing I. The f-ring A is semiprime (prime) if $\{0\}$ is a semiprime (prime) ℓ-ideal . The ℓ-ideal I is *pseudoprime* if $ab = 0$ implies $a \in I$ or $b \in I$. In a semiprime f-ring , a pseudoprime ℓ-ideal contains a prime ℓ-ideal as shown in [**10**, 2.1]. An ideal I of a commutative f-ring is *primary* if $ab \in I$, $a \notin I$ implies $b^n \in I$ for some n. We let $I : J$ denote $\{a \in A : aJ \subseteq I\}$.

Now let X be a topological space, $C(X)$ be the f-ring of all continuous real-valued functions on X with coordinatewise operations and I be an ℓ-ideal of $C(X)$. Williams shows in [**11**, 2.8] that for any ℓ-ideal I, both $I : \sqrt{I}$ and $\langle I\sqrt{I}\rangle$ are intersections of primary ℓ-ideals . He also shows:

Theorem 1 ([**11**, 2.10]) *For a pseudoprime ℓ-ideal I of $C(X)$, the following are equivalent :*
(1) I is primary
(2) $I = I : \sqrt{I}$ or $I = I\sqrt{I}$
(3) $I^2 = I(I : \sqrt{I})$.

Since every primary ℓ-ideal in $C(X)$ is known to be pseudoprime ([**3**, 4.1]), this result characterizes all primary ℓ-ideals in $C(X)$. Later it was shown that one of the implications of this result generalizes to commutative semiprime f-rings with identity element.

Theorem 2 ([**8**, 3.6]) *Let A be a commutative semiprime f-ring with identity element. If an ℓ-ideal I of A satisfies $I = I : \sqrt{I}$ or $I = \langle I\sqrt{I}\rangle$ then I is an intersection*

J. Martinez (ed.), Ordered Algebraic Structures, 181–186.

of primary ℓ-ideals and if I is pseudoprime and satisfies $I = I : \sqrt{I}$ or $I = \langle I\sqrt{I}\rangle$ then I is a primary ℓ-ideal .

In general, this does not characterize primary ℓ-ideals in the class of pseudoprime ℓ-ideals . We will consider what conditions on an f-ring are necessary and sufficient for conditions (2) and (3) of Theorem 1 to characterize primary ℓ-ideals in the class of pseudoprime ℓ-ideals . Primary ℓ-ideals are known to be pseudoprime in several classes of f-rings including uniformly complete f-rings with identity element and commutative semiprime normal f-rings with identity element ([9, 2.6]). So knowing when conditions (2) and (3) of Theorem 1 characterize primary ℓ-ideals in the class of pseudoprime ℓ-ideals will allow characterizations of all primary ℓ-ideals in some classes of f-rings .

An f-ring A with identity element is said to satisfy the *bounded inversion property* if $a \geq 1$ in A implies $a^{-1} \in A$. There is a class of primary ℓ-ideals in a commutative semiprime f-ring satisfying the bounded inversion property that serves as a useful tool in describing when conditions (2) and (3) of Theorem 1 characterize primary ℓ-ideals in the class of pseudoprime ℓ-ideals . The following theorem describes this set of ℓ-ideals . Its result was shown by Gillman and Kohls for $C(X)$ ([3, 4.6]) and remains valid in the more general context given next. Note that if A is an f-ring and I an ℓ-ideal of A, then $I(a)$ will denote the coset of A/I which contains the element a.

Lemma 3 *Let A be a commutative semiprime f-ring with identity element which satisfies the bounded inversion property. Let P be a prime ℓ-ideal of A. If a is a positive nonunit of A/P, then*

$$mP|^a = \{ b \in A/P \colon |b|^m < a^{m-1} \ \ \textit{for all } m \in \mathbb{N} \}$$

and

$$mP|_a = \{ b \in A/P \colon |b|^m \leq a^{m+1} \ \ \textit{for some } m \in \mathbb{N} \}$$

are primary ℓ-ideals of A/P, and $a \in mP|^a$, $a \notin mP|_a$.

Henriksen, in [4] calls an ℓ-ideal I of an f-ring A *square dominated* if $I = \{ a \in A : |a| \leq x^2$ for some $x \in A$ such that $x^2 \in I \}$. A prime ℓ-ideal P is easily seen to be square dominated if and only if it satisfies $P = \langle P^2 \rangle$. Of course in any square-root closed f-ring , every (prime) ℓ-ideal is square dominated. We now give a characterization of when condition (2) of Theorem 1 characterizes primary ℓ-ideals in the class of pseudoprime ℓ-ideals .

Theorem 4 *Let A be a commutative semiprime f-ring with identity element which satisfies the bounded inversion property. The following are equivalent :*

(1) $I = I:\sqrt{I}$ or $I = \langle I\sqrt{I}\rangle$ *characterizes primary ℓ-ideals in the class of pseudoprime ℓ-ideals*

(2) *Every prime ℓ-ideal in A is square dominated, and, for every prime ℓ-ideal P and positive nonunit a in A/P, $mP|^a$ is the smallest primary ℓ-ideal in A/P containing a.*

(3) *Every prime ℓ-ideal in A is square dominated, and, for every prime ℓ-ideal P and positive nonunit a in A/P, $mP|_a$ is the largest primary ℓ-ideal in A/P not containing a.*

(4) *Every prime ℓ-ideal in A is square dominated, and, for every minimal prime ℓ-ideal P and positive nonunit a in A/P, $mP|^a$ is the smallest primary ℓ-ideal in A/P containing a.*

(5) *Every prime ℓ-ideal in A is square dominated, and, for every minimal prime ℓ-ideal P and positive nonunit a in A/P, $mP|_a$ is the largest primary ℓ-ideal in A/P not containing a.*

The keys to proving the previous theorem are the facts that if I is a primary ℓ-ideal , then $\sqrt{I}$ will be a square dominated prime ℓ-ideal and a pseudoprime ℓ-ideal which is an intersection of primary ℓ-ideals is itself primary.

Even in a commutative semiprime f-ring with identity element which satisfies the bounded inversion property, and in which prime ℓ-ideals are square dominated, the conditions given in the previous theorem do not automatically hold. We give a brief description of an f-ring in which condition (1) (and therefore the other conditions) of the previous theorem does not hold.

Example 5 Define $h : [0,1] \to \mathbf{R}$ by

$$h(x) = \begin{cases} 0, & \text{if } x = 0; \\ x(\sum_{n=1}^{\infty} \frac{1}{2^n} x^{1/n})^{-1}, & \text{if } x \neq 0. \end{cases}$$

Then $h(x) \in C([0,1])$. Let $A = \{ f \in C([0,1]) : \exists\, x_f > 0$ such that $f(x) = ax^r h^n(x)$ for some $a \in \mathbf{R}, r \in \mathbf{Q}^+, n \in \mathbf{Z}^+$, and for all $x \in [0, x_f], \}$. Then A can be shown to be a commutative semiprime f-ring with identity element which satisfies the bounded inversion property. It can also be shown that prime ℓ-ideals are square dominated in A. Let $I = \{ f \in A : f(x) \leq nx$ for $x \in [0, x_f]$ and some $n \in \mathbf{N}\}$. Then I is a pseudoprime primary ℓ-ideal . Yet one can show that $\langle I\sqrt{I}\rangle \subset I \subset I : \sqrt{I}$. Indeed, $\sqrt{I} = \{ f \in A : f(x) \leq nx^{1/m}$ for $x \in [0, x_f]$ and some $n, m\}$, which implies $\langle I\sqrt{I}\rangle \subseteq \{ f \in A : f(x) \leq nx^{1+1/m}$ for $x \in [0, x_f]$ and some $n, m\} \subset I \subset \{ f \in A : f(x) \leq nh(x)$ for $x \in [0, x_f]$ and some $n\} \subseteq I : \sqrt{I}$.

Let n be a positive integer. In [7] a commutative f-ring A is said to satisfy the nth*-convexity property* if for any $u, v \in A$ such that $v \geq 0$ and $0 \leq u \leq v^n$, there exists a $w \in A$ such that $u = wv$. A generalization of this property for semiprime f-rings will allow us to characterize those f-rings in which $I^2 = I(I:\sqrt{I})$ characterizes primary ℓ-ideals in the class of pseudoprime ℓ-ideals .

Definition 6 *Let p, q be positive integers with $p \geq q$. A commutative semiprime f-ring A satisfies the (p, q)-convexity property if whenever $u, v \geq 0$ with $0 \leq u^q \leq v^p$, there exists a $w \in A$ such that $u = wv$.*

Note that if n, p, q are positive integers with $n = \frac{p}{q}$, this definition is identical to that of the nth-convexity property in a commutative semiprime f-ring . Note also that $C(X)$ satisfies the (p, q)-convexity property for all p, q with $p > q$ ([2, 1D.3]).

Theorem 7 *Let A be a commutative semiprime f-ring with identity element in which the (p, q)-convexity property holds for some p, q with $q \leq p < 2q$. In the class of pseudoprime ℓ-ideals , $I = I:\sqrt{I}$ or $I = \langle I\sqrt{I}\rangle$ characterizes primary ℓ-ideals if and only if $I^2 = I(I:\sqrt{I})$ characterizes primary ℓ-ideals .*

The proof of this theorem relies on Theorem 2.2, the use of square dominated prime ℓ-ideals and the use of primary ℓ-ideals of the form $mP|^a$, $mP|_a$.

The hypothesis that the f-ring satisfy the (p, q)-convexity property for some p, q with $q \leq p < 2q$ cannot be omitted from this theorem. A description of an example verifying this is given next.

Example 8 Let $C = \{x^r y^s : r, s \in \mathbb{Q}\}$ and B be the semigroup ring over C with real coeffficients and lexicographically ordered such that $1 >> y^p >> \cdots >> y^q >> \cdots >> x^r >> \cdots >> x^s >> \cdots$ for all $p < q, r < s$. Then B is a totally ordered commutative prime ring with identity element. Now let $A = \{\frac{a}{b} : a, b \in B$ and $b \geq 1\}$ with the operations inherited from the quotient ring of B. Then A is a totally ordered commutative prime ring with identity element which satisfies the bounded inversion property. It can be shown that prime ℓ-ideals of A are square dominated and that for every minimal prime ℓ-ideal P and positive nonunit a in A/P, $mP|_a$ is the largest primary ℓ-ideal in A/P not containing a. Hence $I = I:\sqrt{I}$ or $I = \langle I\sqrt{I}\rangle$ characterizes primary ℓ-ideals in the class of pseudoprime ℓ-ideals (Theorem 4).

Let $I = \{a \in I : |a| < nx^p y^q$ where $p, q > 1, n \in \mathbb{N}\}$. Then I is a pseudoprime ℓ-ideal . Also, $\sqrt{I} = \{a \in A : |a| < nx^p y^q$, where $p, q > 0, n \in \mathbb{N}\}$ and $I:\sqrt{I} = \{a \in A : |a| < nx^p y^q$, where $p, q \geq 1, n \in \mathbb{N}\}$. So $\langle I(I:\sqrt{I})\rangle = \{a \in A : |a| < nx^p y^q$, where $p, q > 2, n \in \mathbb{N}\} = \langle I^2\rangle$. But I is not primary since $x^{3/2}y^{3/2} \in I$ and yet $x^{3/2} \notin I$, $y^{3/2} \notin \sqrt{I}$. So $I^2 = I(I:\sqrt{I})$ does not characterize primary ℓ-ideals among pseudoprime ℓ-ideals in A.

Finally we give two classes of f-rings in which one or both of the conditions of Theorem 1 characterize all primary ℓ-ideals of an f-ring . The next theorem was first proven by Kohls for $C(X)$ ([6, 2.1]), and the proof remains valid in the more general context given next.

Theorem 9 *Let A be a commutative semiprime f-ring with identity element for which every ℓ-homomorphic prime image satisfies the 1st-convexity property. If P is a prime ℓ-ideal of A and a is a positive nonunit in A/P then $mP|^a$ is the smallest primary ℓ-ideal in A/P containing a.*

Theorem 9 and Theorem 4 imply that in a commutative semiprime f-ring with identity element for which every ℓ-homomorphic prime image satisfies the 1st-convexity property, $I = I\!:\!\sqrt{I}$ or $I = \langle I\sqrt{I}\rangle$ characterizes primary ℓ-ideals in the class of pseudoprime ℓ-ideals . In a commutative semiprime f-ring with identity element in which the 1st-convexity property holds, every ℓ-homomorphic prime image also satisfies the 1st-convexity property ([7, 2.3]) and every primary ℓ-ideal is pseudoprime ([5, 6.1] and [9, 2.6]). So in a commutative semiprime f-ring with identity element in which the 1st-convexity property holds, $I = I\!:\!\sqrt{I}$ or $I = \langle I\sqrt{I}\rangle$ characterizes all primary ℓ-ideals .

Theorem 10 *Let A be a commutative semiprime f-ring with identity element in which every finitely generated ideal is principal. The following are equivalent :*
(1) *$I = I\!:\!\sqrt{I}$ or $I = \langle I\sqrt{I}\rangle$ characterizes primary ℓ-ideals*
(2) *$\langle I^2\rangle = \langle I(I\!:\!\sqrt{I})\rangle$ characterizes primary ℓ-ideals*
(3) *Every prime ℓ-ideal is square dominated.*

Recall that given a commutative f-ring A and a subset S without zero divisors, there is an f-ring A_S , called the localization of A at S, and an embedding $\lambda : A \to A_S$ such that (i) for every $s \in S$, $\lambda(s)$ is invertible in A_S, and (ii) for any ℓ-homomorphism $\phi : A \to B$ mapping A into an f-ring B such that every $\phi(s)$ is invertible, there exists an ℓ-homomorphism $\bar{\phi} : A_S \to B$ such that $\bar{\phi} \circ \lambda = \phi$. Let A be a commutative f-ring with identity element in which every finitely generated ideal is principal. If $S = \{ s \in A : s \geq 1 \}$, then A_S, the localization of A at S satisfies the 1st-convexity condition as shown in [8, 2.8]. Most of the previous theorem can be proven by embedding A into its localization at S and using Theorems 9, 4, and 7.

References

1. A. Bigard, K. Keimel and S. Wolfenstein, *Groupes et anneaux reticules*, Lecture Notes in Mathematics *608* (Springer Verlag, 1977).
2. L. Gillman and M. Jerison, *Rings of continuous functions*, (Springer Verlag, 1960).
3. L. Gillman and C. Kohls, *Convex and pseudoprime ideals in rings of continuous functions*, Math. Zeitschr. *72* (1960), 399 - 409.
4. M. Henriksen, *Semiprime ideals of f-rings*, Symposia Math. *21* (1977), 401 - 409.
5. C. Huijsmans and B. de Pagter, *Ideal theory in f-algebras*, Trans. Amer. Math. Soc., *269* (1982), 225 - 245.
6. C. Kohls, *Primary ideals in rings of continuous functions*, Amer. Math. Monthly, *71* (1964), 980 - 984.
7. S. Larson, *Convexity conditions on f-rings*, Can. J. Math., *38* (1986), 48 - 64.
8. S. Larson, *Minimal convex extensions and intersections of primary ℓ-ideals in f-rings*, J. of Algebra, to appear.
9. S. Larson, *Pseudoprime ℓ-ideals in a class of f-rings*, Proc. Amer. Math. Soc., *104* No. 3 (1988), 685 - 692.
10. H. Subramanian, *ℓ-prime ideals in f-rings*, Bull. Soc. Math. France *95* (1967), 193 - 203.
11. R. D. Williams, *Intersections of primary ideals in rings of continuous functions*, Can. J. Math., *24* (1972), 502 - 519.

EPIMORPHISMS OF f-RINGS

Niels Schwartz

Fakultät für Mathematik und Informatik
Universität Passau
Postfach 2540
8390 Passau
W. Germany

Introduction: A total order of an integral domain A has a unique extension to a total order of the quotient field qf(A) ([10], p. 20, Satz 3). If R is the real closure of qf(A) ([10], Chapter II §2) then the canonical monomorphism A → R is an epimorphism in the category of totally ordered commutative rings ([7]). In the present paper monomorphic epimorphisms between f-rings are investigated. More precisely, let $\mathcal{A}$ be the category of commutative f-rings with 1, $\mathcal{B}\subset\mathcal{A}$ the full subcategory of reduced f-rings. If f:A → B is an epimorphism in $\mathcal{A}$ then f can be written as the composition of a surjective epimorphism and a monomorphic epimorphism. Surjective epimorphisms are well understood. So it remains to study monomorphic epimorphisms. To avoid complications arising from nilpotency we will consider the following question: Given $A\in\mathcal{B}$, what are the monomorphic epimorphisms f:A → B in $\mathcal{B}$? This problem can be solved completely using some real algebra developed in the context of real algebraic geometry. It is shown that there are maximal epimorphic extension. Every epimorphic extension is contained in a maximal epimorphic extension. Given $A\in\mathcal{B}$ we let the epimorphic extensions of A in $\mathcal{A}$ be the objects of a category $\mathcal{C}$. A morphism h:(f:A → B) → (g:A → C) in $\mathcal{C}$ is a homomorphism h:B → C such that g=hf. In general there is no largest epimorphic extension of A, but the category always contains a final object.

I wish to thank J. Madden for bringing this problem to my attention (also see his contributions to Problem Session IV, these Proceedings). Although the answer of the question is not very hard for someone familiar with the methods used in real algebraic geometry, the

J. Martinez (ed.), Ordered Algebraic Structures, 187–195.

solution is presented here to advertise a closer cooperation between the groups working in real algebraic geometry and in the theory of lattice ordered rings. A short glance at Chapter 10 of [1] and at Chapter I of [11] should convince everyone that there are many connections between these two fields.

1. A characterization of epimorphic extensions. Let A be a commutative reduced f-ring ([1], Definition 9.1.1). Thus, A is a lattice-ordered subring of a direct product of totally ordered fields with componentwise lattice order. Spec(A) is the prime spectrum of the ring A. Sper(A) denotes the real spectrum of A ([2]; [4]; [8]; [11]). The subsets $D(a)=\{\alpha|a\notin -\alpha\}$ $(a\in A)$ of Sper(A) form a subbasis of a topology ([4]) which is called the Zariski topology of Sper(A). A subset of Sper(A) is constructible if it belongs to the Boolean algebra of subsets generated by the D(a). The constructible subsets of Sper(A) are a basis of the constructible topology ([5], I 7.2.11) which is also called the patch topology by Hochster ([6]). The subsets of Sper(A) which are closed in the constructible topology are called pro-constructible ([5], I §7.2; [11], Chapter I §1). If $Z\subset Sper(A)$ is pro-constructible then Z is a spectral space ([6]). Z_{min} denotes the set of generic points of Z ([5], 0 2.1.1). For $z\in Z$, $\rho(z)$ is the real closed residue field of z ([11], Chapter I, §1). We set

$$Z(A)=\{\alpha\in Sper|A^{+}\subset\alpha\}$$

(with $A^{+}\subset A$ the positive cone). This is a pro-constructible subset of Sper(A). Also,

$$T(A)=\{supp(\alpha)|\alpha\in Z(A)\}$$

(with supp:Sper(A) → Spec(A) the support function - see [2], Proposition 4.3.2; [8], Definition 3.1, [11], Chapter I, §1) is a pro-constructible subset of Spec(A). In fact, T(A) is the set of prime ideals of A which are ℓ-ideals. The support function supp:$Z(A) \to T(A)$ is an isomorphism of spectral spaces (i.e., it is continuous in both the constructible topology and the Zariski topology). With these notations we will prove:

<u>Theorem</u> <u>1</u>: Let $f:A \to B$ be a morphism in $\mathcal{B}$. f is an epimorphic extension in $\mathcal{B}$, if and only if the following conditions hold:

(i) $Z(A)_{min} \subset \mathrm{Sper}(f)(Z(B))$.

(ii) The restriction $Z(B) \to Z(A)$ of Sper(f) is injective.

(iii) For $z \in Z(B)$, $t = \mathrm{Sper}(f)(z) \in Z(A)$ the homomorphism $\rho(t) \to \rho(z)$ of residue fields is an isomorphism.

<u>Proof</u>: First suppose that f is an epimorphic extension in $\mathcal{A}$. Then every minimal prime ideal $P \subset A$ extends to a minimal prime ideal $Q \subset B$ ([3], p. 96, Proposition 16). The sets of minimal prime ideals of A and B are exactly $T(A)_{min}$ and $T(B)_{min}$ ([1], Theoreme 9.3.2). By the correspondence between T(A) and Z(A) (resp., T(B) and Z(B)), condition (i) holds. Now suppose that (ii) is false. Pick $u, v \in Z(B)$, $u \neq v$ with $\mathrm{Sper}(f)(u) = z = \mathrm{Sper}(f)(v)$. By the amalgation property of real closed fields ([9], Satz 3.22, Satz 4.7) there is a commutative diagram

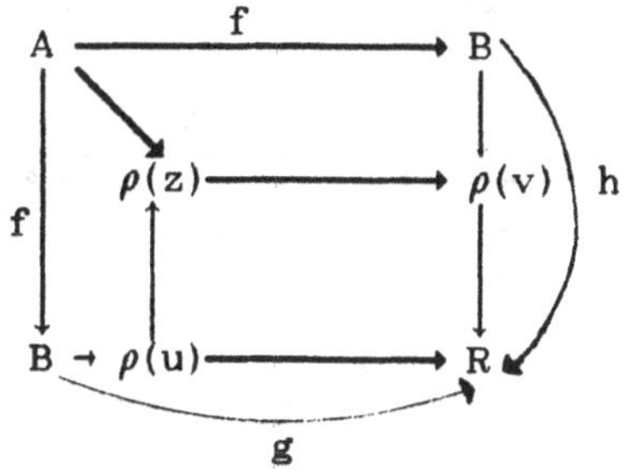

where R is an appropriate real closed field. From $u \neq v$ it follows that $g \neq h$. But $gf = hf$ implies $g = h$ since f is an epimorphism, a contradiction. Finally suppose that there is some $z \in Z(B)$ such that $\rho(t) \to \rho(z)$ (with $t = \mathrm{Sper}(f)(z)$) is not surjective. Then there is a commutative square

$$\begin{array}{ccc} \rho(t) & \longrightarrow & \rho(z) \\ \downarrow & & \downarrow{\scriptstyle h} \\ \rho(Z) & \xrightarrow[g]{} & R \end{array}$$

of real closed fields such that $g \neq h$. This leads to a diagram as above and to the same kind of contradiction.

Conversely, assume that (i), (ii), (iii) hold for f. From (i) (in connection with [1], Théoréme 9.3.2) it is immediately clear that every minimal prime ideal of A extends to a prime ideal of B. Since A is reduced this means that f is injective. Now pick a reduced f-ring C and homomorphisms $g,h:B \to C$ such that $gf=hf$. We wish to show that $g=h$. For this it suffices to prove that $\pi_I g=\pi_I h$ for every canonical epimorphism $\pi_I:C \to C/I$, $I \subset C$ a minimal irreducible ℓ-ideal, i.e., a minimal prime ideal ([11], Théoréme 9.3.2). So we may assume that C is a totally ordered domain. Let $w \in Z(C)$ be the generic point and set $u=\mathrm{Sper}(g)(w)$, $v=\mathrm{Sper}(h)(w)$. From $gf=hf$ it follows that $\mathrm{Sper}(f)(u)=z=\mathrm{Sper}(f)(v)$. By condition (ii), this implies $u=v$ and condition (iii) now shows that $\rho(z) \to \rho(u)$ is an isomorphism. This gives us a commutative diagram

$$\begin{array}{ccccc} A & \xrightarrow{f} & B & \overset{g}{\rightrightarrows} & C \\ \downarrow & & \downarrow & & \downarrow{\scriptstyle c} \\ \rho(z) & \xrightarrow{\cong} & \rho(u) & \underset{\bar{h}}{\overset{\bar{g}}{\rightrightarrows}} & \rho(w). \end{array}$$

The equality of gf and hf implies that $\bar{g} = \bar{h}$, i.e.,

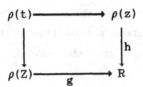

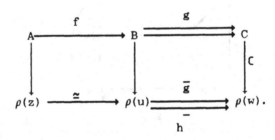

$$B \longrightarrow \rho(u) \overset{\bar{g}}{\underset{\bar{h}}{\rightrightarrows}} \rho(w)$$

agree. Since C is a subring of $\rho(w)$, we see that g=h. □

With the methods used in the proof of Theorem 1 it can also be proved that the categories $\mathcal{A}$ and $\mathcal{B}$ both have the amalgation property.

2. A class of epimorphic extensions. Continuing with the notation of section 1 we let $Z \subset Z(A)$ be a pro-constructible subset containing $Z(A)_{min}$. Then the support function

$$\text{supp}: Z \longrightarrow T = \text{supp}(Z)$$

is an isomorphism of spectral spaces again. T is a pro-constructible set of prime ℓ-ideals of A containing the minimal prime ideals of A. Since A is reduced, the canonical homomorphisms

$$A \longrightarrow \prod_{t \in T} A/t \longrightarrow \prod_{z \in Z} \rho(z)$$

are monomorphisms of f-rings. We define $A_Z \subset \prod_{z \in Z} \rho(z)$ to be the ring of constructible sections on Z over A ([11], p. 8). Let $i_Z: A \to A_Z$ be the canonical monomorphism. The elements of A_Z can be represented in the following way: Given $a \in A_Z$ there is a constructible partition $Z = Z_1 \cup \ldots \cup Z_n$ and for each i there is an abstract semi-algebraic function a_i on Z_i over A ([11], Chapter 1, §2) such that $a|Z_i = a_i$ for all i.

Without proof we record the following properties of A_Z:

Proposition: (a) A_Z is a real closed ring ([11], Definition I 4.1).

(b) There is a canonical homeomorphism $Z \to \mathrm{Sper}(A_Z)$ (Z with the constructible topology) associating with $t \in Z$ the point of $\mathrm{Sper}(A_Z)$ given by $A_Z \to \prod_{z \in Z} \rho(z) \to \rho(t)$.

(c) For $z \in Z$, $\{a \in A_Z | a(z)=0\}$ is a maximal ideal with residue field $\rho(z)$.

(d) Every ideal of A_Z is radical.

(e) $i_Z : A \to A_Z$ is a monomorphism of f-rings.

Using this information about A_Z we can prove

Theorem 2: The canonical homomorphism $i_Z : A \to A_Z$ is an epimorphic extension in $\mathcal{A}$.

Proof: Since both A and A_Z are reduced and conditions (i) - (iii) of Theorem 1 hold, i_Z is an epimorphic extension in $\mathcal{B}$. It remains to show that the same is true in $\mathcal{A}$. To prove this it suffices to consider some totally ordered $C \in \mathcal{A}$ and homomorphisms $g,h : A_Z \to C$ such that $gi_Z = hi_Z$. The nilradical of C is a convex prime ideal ([1], Proposition 9.2.3) and the natural epimorphism $p : C \to C_{red}$ is a homomorphism of f-rings. Since i_Z is epimorphic in the category $\mathcal{B}$, $pgi_Z = phi_Z$ implies $pg = ph$. Since every ideal of A_Z is radical, $\ker(pg) = \ker(g)$, $\mathrm{Ker}(ph) = \ker(h)$. So g and h define the same point z in the real spectrum of A_Z. This gives a diagram

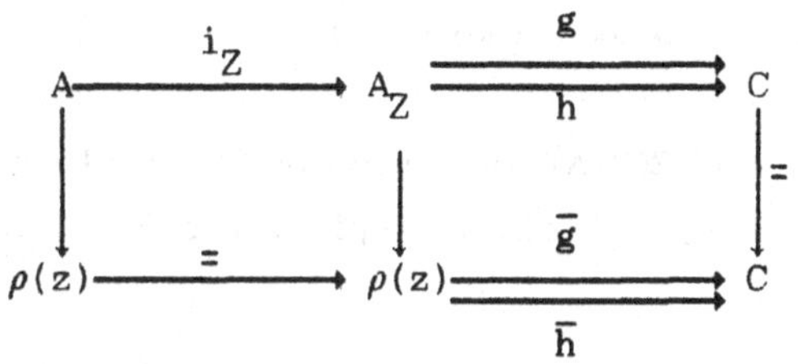

in which $A \to \rho(z)$ is an epimorphism of totally ordered rings ([7]). This implies $\bar{g} = \bar{h}$ and, consequently, $g=h$. □

3. <u>The classification of monomorphic epimorphisms</u>. As a consequence of the previous sections we can now determine all $\mathcal{A}$-epimorphic extensions $i:A \to B$ with $A,B \in \mathcal{B}$. So, let such an extension be given. By Theorem 1, $Sper(i):Z(B) \to Z(A)$ is injective and $Z=Sper(i)(Z(B)) \supset Z(A)_{min}$. Since $Sper(i)$ is a morphism of spectral spaces, $Z \subset Z(A)$ is a pro-constructible subset. By condition (iii) of Theorem 1 both A_Z and B are subrings of $\prod_{z \in Z} \rho(z)$. We clearly have $A_Z \subset B_{Z(B)} \subset \prod_{z \in Z} \rho(z)$. If we assume that there is some $b \in B_{Z(B)} \setminus A_Z$ then the constructible topology on $Z(B)$ must be finer than the constructible topology on Z. Since both spaces are compact, this is a contradiction. So we see that $B \subset B_{Z(B)} = A_Z$. This proves

<u>Theorem</u> <u>3</u>: Let $i:A \to B$ be a monomorphism in $\mathcal{B}$. i is an epimorphic extension in $\mathcal{B}$ if and only if $Sper(i):Z(B) \to Z(A)$ is injective and (setting $Z=im(Sper(i))$) there is a commutative diagram

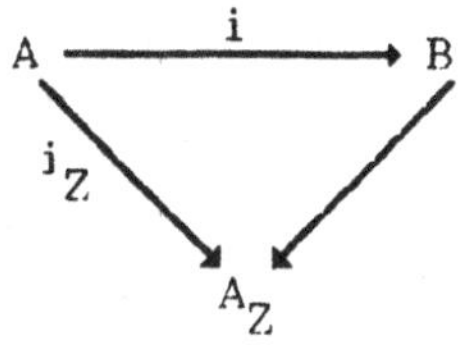

As an immediate consequence we note:

Corollary: If $A\in\mathcal{B}$ and $Z\subset Z(A)$ is a pro-constructible subset containing $Z(A)_{min}$ then $i_Z:A \to A_Z$ is a maximal epimorphic extension in $\mathcal{B}$.

In general there are many possible choices for the set Z. So, one cannot expect to have a largest epimorphic extension of $A\in\mathcal{B}$. However, there is a smallest pro-constructible subset $Z_o\subset Z(A)$ containing $Z(A)_{min}$. This is the closure of $Z(A)_{min}$ in the constructible topology. We will use Z_o now to determine the final object in the category $\mathcal{C}$ (see the Introduction). By Theorem 2, $i_o=i_{Z_o}:A \to A_o=A_{Z_o}$ is an object of $\mathcal{C}$. Now let $i:A \to B$ be any object of $\mathcal{C}$. If $p:B \to B_{red}$ is canonical then pi is a morphism in $\mathcal{B}$ ([1], Théoréme 9.2.6) and an object of $\mathcal{C}$ (since $i(A)\cap Nil(B)=\{0\}$). If we let $Z=im(Sper(pi))\subset Z(A)$ then $Z\subset Z(A)$ is a pro-constructible subset containing Z_o. By Theorem 3 we may consider B_{red} as a subring of A_Z. Finally, composing all these homomorphisms with the canonical restriction homomorphism $A_Z \to A_o$ we obtain the diagram

$$\begin{array}{ccccc} & & A & & \\ & \swarrow & & \searrow & \\ B \to B_{red} & & \subset & & A_Z \to A_o. \end{array}$$

This proves

Theorem 4: $i_o:A \to A_o$ is the final object in the category $\mathcal{C}$.

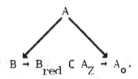

References:

1. A Bigard, K. Keimel, S. Wolfenstein: Groupes et Anneaux Réticules. Lecture Notes in Mathematics, vol. 608, Springer 1977

2. J. Bochnak, M. Coste, M.-F. Roy: Géométrie algébrique réelle. Ergebnisse der Mathematik und ihrer Grenzgebiete, 3. Folge, Bd. 12, Springer 1987

3. N. Bourbaki: Algébre Commutative. Chapitre I, II Hermann 1961

4. M. Coste, M.-F. Roy: La topologie du spectre reel. In: Ordered Fields and Real Algebraic Geometry (D.W. Dubois, T. Recio eds.), Contemporary Mathematics, vol. 8, 1982

5. A. Grothendieck, J.A. Dieudonné: Eléments de Géométrie Algébrique I. Grundlehren der mathematischen Wissenschaften, Springer 1971

6. M. Hochster: Prime ideal structure in commutative rings. Trans. AMS 142, 43-60 (1969)

7. J. Isbell: Notes on ordered rings. Alg. Univ. 1, 393-399 (1972)

8. T.Y. Lam: An introduction to real algebra. Rocky Montain J. Math. 14, 767-814 (1984)

9. A. Prestel: Einführung in die mathematische Logik und Modelltheorie. Vieweg Studium-Aufbaukurs Mathematik, Vieweg 1986

10. S. Prieß-Crampe: Angeordnete Strukturen: Gruppen, Körper, projektive Ebenen. Ergebnisse der Mathematik und ihrer Grenzgebiete, vol. 98, Springer 1983

11. N. Schwartz: The Basic Theory of Real Closed Spaces. Memoirs of the AMS, No. 397, 1989

Representation of a real polynomial f(X) as a sum of 2m-th powers of rational functions[1)]

A.Prestel
Universität Konstanz
Fakultät für Mathematik
Postfach 5560
7750 Konstanz

M. Bradley
Universidad Complutense de Madrid
Facultad de Ciencias Matematicas
Depto. de Algebra y Fundamentos
E-28040 Madrid

1. Introduction and result

From Becker's Satz 2.14 in [B_1] it follows that a polynomial $f \in \mathbb{R}[X]$ admits a representation

(1) $$f = \sum_{i=1}^{\sigma} \frac{g_i^{2m}}{h^{2m}} \quad \text{with } g_i, h \in \mathbb{R}[X]$$

if and only if f satisfies the following three conditions:

(2) (i) 2m divides deg f
(ii) 2m divides the order of every real zero of f
(iii) f is positive semidefinite

Once f satisfies these conditions, the problem arises how to obtain a representation (1) for f . This paper is concerned with that problem.

From [B_2] it is known that there is a bound σ in (1), depending on m , which works uniformly for all $f \in \mathbb{R}[X]$ satisfying (2),(i)-(iii). In case m=2, Becker has first shown that $\sigma \leq 36$ and Schmid has recently improved this to $\sigma \leq 24$ (see [S]).

[1)] The result of this paper was obtained when the first author was working on her thesis [Br] under the supervision of the second author.

J. Martinez (ed.), Ordered Algebraic Structures, 197–207.

Fixing a bound σ and the degree of f in (1), the second author has shown for m>1 in [P_2] that the dependence of g_i and h on the coefficients of f is rather complicated. Surprisingly, even the degree of h (and hence of the g_i) depends on the coefficients of f and not just on the degree of f [2]. In fact it was shown in [P_2] that in a representation (1) of the polynomial

(3) $X^4 + nX^2 + 1$,

the degree of h tends to infinity as n does. The proof of this result was obtained using non-standard methods and neither gives a constructive lower bound of deg h depending on n , nor does it give any idea on what the degree of h actually depends.

However, using Becker's Satz 2.14 from [B_1] not just for representations (1) over $\mathbb{R}$ but also over any real closed field, the Compactness Theorem from Model Theory enables us to prove the following

THEOREM A <u>There exists a computable function</u> $\delta:\mathbb{N}^3 \to \mathbb{N}$ <u>such that for all polynomials</u> $f=X^d+a_{d-1}X^{d-1}+\ldots+a_0 \in \mathbb{R}[X]$ <u>with</u> $2m|d$, $|a_i|<N$ <u>for all</u> $1\le i\le d-1$, <u>and such that every monic polynomial</u> g of degree d <u>with</u> $\|f-g\|<\frac{1}{M}$ <u>is strictly positive</u>, f <u>admits a representation</u> (1) <u>with</u> deg h $\le \delta(d,N,M)$.

Here $\|h\|$ is the absolute value of the coefficients of h, and h being strictly positive means that $h(x)>0$ for all $x\in\mathbb{R}$.

The bound σ of a representation (1) in Theorem A may be any σ such that every sum of 2m-th powers from the field R(X) with R being an arbitrary real closed field, can be written as a sum of σ many 2m-th powers from R(X). The bound $\sigma=24$ from [S] e.g. satisfies this condition. Thus in the following we always may assume σ to be 24.

---

[2] In case m=1, the degree of h always may be chosen to be zero (see discussion at the end of this paper).

Since every positive semidefinite polynomial f may be uniquely written in the form

$$f = a(X-\alpha_1)\cdot\ldots(X-\alpha_s)\cdot f^*$$

where $a,\alpha_1,\ldots,\alpha_s \in \mathbb{R}$, $a>0$ and f^* is monic and strictly positive, we see from (2) and Theorem A that the degree of h in a representation (1) of f depends only on deg f^*, the 'size' of the coefficients of f^* and the 'quality' of f^* of being strictly positive. In our discussion at the end of this paper we shall make this more precise.

2. Proof of the theorem

For each $a=(a_0,\ldots,a_{d-1}) \in \mathbb{R}^d$ we consider the polynomial

$$f(a,X) = X^d + a_{d-1}X^{d-1}+\ldots+a_0 .$$

Let S be a subset of $\mathbb{R}^d$. If this set is given by some formula $\varphi=\varphi(x_0,\ldots,x_{d-1})$ in the language of ordered fields, we may as well consider the set

$$S_R=\{a \in R^d \mid a \text{ satisfies } \varphi \text{ in } R\}$$

over an arbitrary real closed field R and to every $a \in S_R$ the corresponding polynomial $f(a,X) \in R[X]$.

The following lemma is then an important consequence of the Compactness Theorem.

LEMMA <u>Assume that for every</u> $a \in S$ <u>the polynomial</u> $f(a,X)$ <u>admits a representation</u> (1). <u>Then there is a bound</u> δ <u>such that, for every</u> $a \in S$: deg $h \leq \delta$ <u>in some representation</u> (1) <u>of</u> $f(a,X)$ <u>if and only if for each real closed field</u> R <u>and each</u> $a \in S_R$, <u>the polynomial</u> $f(a,X)$ <u>admits a representation</u> (1) <u>with</u> $g_i,h \in R[X]$. <u>The bound</u> δ <u>can be chosen to depend recursively on the formula</u> φ .

Proof: It is easy to see that representability (1) for a polynomial f(a,X) can be expressed by a countable disjunction of formulas $\varphi_t(a)$ where, for each $t \in \mathbb{N}$, the formula $\varphi_t(a)$ expresses the representability with deg $h \leq t$ and hence deg $g_i \leq t + \frac{d}{2m}$. The existence of a bound on deg h for all $a \in S$ is thus equivalent to the condition that for all possible coefficients $a \in S$, the infinite disjunction

$$\bigvee_{t=1}^{\infty} \varphi_t(a)$$

may be replaced by a finite subdisjunction

$$\bigvee_{t=1}^{\delta} \varphi_t(a)$$

for some $\delta \in \mathbb{N}$.

Hence, if there is a bound δ for S , the implication

$$(4) \qquad \forall a(\varphi(a) \rightarrow \bigvee_{t=1}^{\delta} \varphi_t(a))$$

holds in $\mathbb{R}$ and thus, by the Tarski-Principle, holds in all real closed fields R. Therefore, every f(a,X) with $a \in S_R$ admits a representation (1).

Conversely, if the polynomial f(a,X) admits a representation (1) for all real closed fields R and all $a \in S_R$, we have that the implication

$$(5) \qquad \forall a(\varphi(a) \rightarrow \bigvee_{t=1}^{\infty} \varphi_t(a))$$

holds in all real closed fields R . Now the Compactness Theorem for Model Theory tells us (compare e.g.[P_3]) that there is some $\delta \in \mathbb{N}$ such that implication (4) already holds in all real closed fields R . In particular, (4) holds in $\mathbb{R}$.

Now let δ_S be the smallest bound of deg h which works for all $a \in S$. This can be expressed by

$$(6) \qquad \forall a(\varphi(a) \rightarrow \varphi_{\delta_s}(a)) \wedge \exists a(\varphi(a) \wedge \neg \varphi_{\delta_s - 1}(a)).$$

Since formula (6) is true in $\mathbb{R}$, by the completeness of the theory of real closed fields, it is derivable from the axioms of this theory. Since all formulas derivable from those axioms can be recursively produced (enumerated), we can recursively obtain the value δ_S depending only on the formula φ.

The crucial point in applying the Compactness Theorem in the above proof is that the implication (5) holds not just in $\mathbb{R}$, but in an axiomatizable class of fields.

In order to prove Theorem A, we will first convince ourselves that the conditions on the coefficients $a_0,\dots,a_{d-1}$ in Theorem A can be expressed by some formula $\varphi(a_0,\dots,a_{d-1})$ in the language of ordered fields. Given $d,N,M \in \mathbb{N}$, the conditions are expressed by

(7) $$\bigwedge_{i=0}^{d-1} |a_i|<N \wedge \forall b_0\dots b_{d-1}\Big(\bigwedge_{i=0}^{d-1} |a_i - b_i|<\frac{1}{M} \;\rightarrow\; \forall x \;\; x^d+\dots+b_0>0\Big)$$

We shall now show for every real closed field R and all $a \in R^d$ satisfying $\varphi(a)$ in R , that the poynomial $f(a,X)$ admits a representation (1). By Becker's Satz 2.14 from [B_1] (and its proof), it suffices to show that for every valuation v of the field $F=R(X)$ such that the residue field $\bar{F}_v$ is a subfield of $\mathbb{R}$, the value $v(f)$ is divisible by 2m.

If R is archimedean ordered, this is equivalent to condition (2), (i) and (ii), since in that case v must be trivial on R . From the assumption that 2m|d and f being strictly positive, this is guaranteed.

Now assume that R is non-archimedean ordered. If we restrict the valuation v to R, it is no longer trivial.

In fact, since the residue field $\bar{R}$ is a subfield of $\mathbb{R}$, the valuation ring A of this restriction is the convex hull of $\mathbb{Q}$ in R. Therefore, the condition $|a_i| < N$ implies that all coefficients lie in A. Thus, we may consider the residue polynomial $f(\bar{a},X) \in \bar{R}[X]$:

$$f(\bar{a},X) = X^d + \bar{a}_{d-1}X^{d-1}+\dots+\bar{a}_0 \;.$$

From the theory of real places (see[P_1], §7 and §8) we know that $\overline{R}$ is also real closed and the valuation $v_{|R}$ is henselian. Thus in particular, there is an embedding $\tau:\overline{R}\to R$ such that for all $a\in A$

$$\overline{\tau(\bar{a})} = \bar{a} \ .$$

This means that $a-\tau(\bar{a})$ lies in the maximal ideal of A . Hence

$$|a-\tau(\bar{a})| < r$$

for all $r\in\mathbf{Q}$, $r>0$. If we now put

$$g(X) = X^d + \tau(\bar{a}_{d-1})X^{d-1}+\ldots+\tau(\bar{a}_0)$$

we see that

$$\|f-g\| < \frac{1}{M}$$

and therefore g has to be strictly positive on R .

Now it is easy to prove that $2m|v(f)$. We shall distinguish two cases:

Case 1: $v(x)<0$. In this case it is clear that $v(f)=v(X^d)=dv(X)$ and thus $2m|v(f)$.

Case 2: $v(X)\geq 0$. This clearly implies $v(f)\geq 0$. If $v(f)=0$ we are done. In case $v(f)>0$ we would obtain in the residue field $\overline{F}$ that $f(\bar{a},\overline{X})=0$. Since F is an ordered extension of R, the element $\overline{X}\in\overline{F}$ must already belong to $\overline{R}$. But then we get $g(\tau(\overline{X}))=0$ through the embedding τ , contradicting the strict definiteness of g.

Thus we have proved (using the Lemma) the existence of a bound δ_S depending effectively on φ . Since the formula φ depends effectively on d,N,M, the function

$$\delta:\mathbb{N}^3\to\mathbb{N}$$

defined by $\delta(d,N,M)=\delta_S$ is obviously recursive.

3. Discussion of the result.

Let $f \in \mathbb{R}[X_1,\ldots,X_n]$ be a polynomial which is positive semidefinite, i.e. $f(a_1,\ldots,a_n) \geq 0$ for all $a_i \in \mathbb{R}$. Then by Artin's solution of Hilbert's 17th Problem, f has a representation

$$f = \sum_{i=1}^{\sigma} \frac{g_i^2}{h^2}.$$

By a result of Pfister, σ may be taken as 2^n. In case n=1 it is easily seen that f is already a sum of two polynomial squares. Thus in this case no non-constant denominator is needed. For n>1 it is also well-known that there is a bound on deg h depending recursively on n and the degree of f. This fact follows from the above Lemma since Artin's solution (and Pfister's Theorem) hold over all real closed fields R, i.e. f is positive semi-definite over R if and only if f has a representation as a sum (of 2^n) squares of rational functions over R.

As we can see from $[P_2]$, the situation is much more complicated for m>1. Even for n=1 there is no bound on deg h depending only on deg f. As Theorem A shows, some properties of the coefficients are also involved. In fact, if we write a positive semidefinite polynomial $f \in \mathbb{R}[X]$ as a product

$$(8) \qquad f = a(X-\alpha_1)\ldots(X-\alpha_s)\, f^*$$

with $a, \alpha_1, \ldots, \alpha_s \in \mathbb{R}$ and $a>0$, f^* monic and strictly positive, and if we assume that f has a representation (1), then deg h depends only on $d^* = \deg f^*$, the size of the coefficients of f^*, and the 'quality' of f^* of being strictly positive, i.e. on the size of a neighbourhood $U_{\frac{1}{M}}(f) = \{g \mid \|f-g\| < \frac{1}{M}\}$ such that $g \in U_{\frac{1}{M}}(f)$ is still strictly positive.

We have shown that the dependence on the parameters (d^*, N, M) of f^* is recursive. Thus the question arises of how to obtain effectively these parameters of f^* from f. If the coefficients $a_0, \ldots, a_{d-1}$ of f are given effectively (say e.g. are rational numbers or computable real numbers), how can we recursively obtain d^*, N and M? Since for a given f , the numbers N and M are not uniquely determined, we may fix them by requiring that they should be minimal, satisfying the conditions of Theorem A for f^*.

One way to obtain those parameters would simply be to consider the formula

$$\varphi_{d*,N,M} = \varphi_{d*,N,M}(a_0,\dots a_{d-1})$$

over the reals which expresses the following fact:

"f has a factorization (8) with f^* satisfying (7) for d^*, N,M where N und M are minimal".

Exactly one formula $\varphi_{d*,N,M}$ is true in $\mathbb{R}$ and, thus, the decidability of the theory of $\mathbb{R}$ allows us to find this formula (and, hence, d^*, N and M) effectively.

As we can see from the above considerations, the main problem for obtaining d^*, N and M is that we should have a factorization of f , i.e. that we should know its zeros. Knowing the zeros of f , however, there is in fact a more satisfactory version of our theorem:

THEOREM B <u>There exists a computable function</u> $\delta : \mathbb{N}^3 \to \mathbb{N}$ <u>such that for all</u>

$$f = \prod_{j=1}^{d} (X-(\alpha_j+\beta_j\sqrt{-1})) \in \mathbb{R}[X]$$

<u>with</u> $\alpha_j, \beta_j \in \mathbb{R}$ <u>the polynomial</u> f <u>admits a representation</u> (1) <u>with</u> $\deg h \le \delta(d,N,M)$ <u>if we have</u> 2m|d , 2m <u>divides the multiplicity of the real roots</u> α_j (i.e. $\beta_j = 0$) <u>and</u>

(i) $|\alpha_j|, |\beta_j| \le N$

(ii) $\beta_j = 0$ or $\frac{1}{M} < |\beta_j|$

<u>for all</u> $1 \le j \le d$.

From (i) it follows that all coefficients of f are bounded by some rational number. From (ii) we see that the non-real zeros of f do not come to close to the x-axis.

The proof of this theorem is essentially the same as that of Theorem A . We have to show that the conditions of Theorem B can be expressed by some first order formula

$$\varphi = \varphi(\alpha_1,\dots,\alpha_d,\beta_1,\dots,\beta_d)$$

and that every polynomial

$$f(\alpha,\beta,X) = \prod_{j=1}^{d}(X-(\alpha_j+\beta_j\sqrt{-1})) \in R[X]$$

satisfying φ over an arbitrary real closed field R admits a representation (1) over R.

Let $2m|d$. Now the formula φ may be taken as the conjunction of

$$\bigwedge_{j=1}^{d} (|\alpha_j| \le N \wedge |\beta_j| \le N \wedge (\beta_j = 0 \vee \tfrac{1}{M} < |\beta_j|))$$

and a finite disjunction φ', giving all the possibilities for the α_j and β_j such that 2m divides the multiplicity of the real roots and such that the non-real roots split into pairs which are conjugated.

Now, if $\varphi(\alpha,\beta)$ holds in some real closed field R , the polynomial $f=f(\alpha,\beta,X)$ factors into

$$f = g^{2m} f^*$$

where g has only real roots and f^* has no real root. Thus, if v is as in the proof of Theorem A , it remains to show that $2m|v(f)$. From the conditions of Theorem B it follows that the residue polynomial $f^*(\bar{\alpha},\bar{\beta},X)$ has no root in the real closed field $\bar{R}$. Thus, either $v(X)<0$ and $v(f)=v(x^d)$ or $v(X)\ge 0$ and $v(f^*)=0$. In both cases we have $2m|v(f)$. This proves Theorem B .

Now that we know the dependence of deg h in a representation (1), it would be interesting to get more information about the coefficients of h and the g_i for fixed d,N and M. For instance, it would be interesting to know whether it is possible to choose the coefficients in such a way that they would depend continuously on the coefficients of f . According to Delzell's result (see[D]), this is possible for m=1 , even for positive definite polynomials f in n variables $X_1,\ldots,X_n$.

Similar theorems as Theorem A (or B) of this paper have meanwhile been obtained by the second author for the case m>1 and polynomials f in two variables X_1,X_2. This will be published in a forthcoming paper. In case the polynomial f has more than two variables, necessary and sufficient conditions for the existence of a representation (1) are known (see [P_2]). But nothing is known in this case about the dependence of deg h.

References

[B_1] Becker, E.: Summen n-ter Potenzen in Körpern. J. reine angew. Math. 307/308 (1979), 8-30

[B_2] Becker, E.: The real holomorphy ring and sums of 2n-th powers. Lecture Notes in Math. 959 (Springer, 1982), 139-181

[Br] Bradley, M.: Aspectos cuantitativos y cualitativos finitistas en summas de potencias 2m-esimas de polinomios. Thesis. Santander 1987

[D] Delzell, Ch.: A sup-inf-polynomially varying solution to Hilbert's 17th Problem. (Preprint)

[P_1] Prestel, A.: Lectures on formally real fields. Lecture Notes in Math. 1093 (Springer, 1984)

[P_2] Prestel, A.: Model theory of fields: An application to positive semidefinite polynomials. Soc.Math. de France, 20 sèrie, mèmoire 16 (1984), 53-65

[P_3] Prestel, A.: Model theory applied to some questions about polynomials. Contr. to General Algebra 5. (Teubner 1986)

[S] Schmid, J.: Eine Bemerkung zu den höheren Pythagoraszahlen reeller Körper. man.math. 61 (1988), 195-202

PARTIALLY ORDERED $*$-DIVISION RINGS

John Dauns
Department of Mathematics
Tulane University
New Orleans, LA 70118

INTRODUCTION

We start with a skew field K with an involution "$*$" and having a partial order of a certain kind (po - $*$ - sfield K). It is assumed that K satisfies the (SRP) (square root for positive elements) axiom, which roughly says that every positive symmetric element of K has a symmetric square root in K which is unique up to ± 1. Then we show how all of these can be extended to a skew Laurent power series extension division ring $K((x;\delta))$ of K. I.e. $K((x;\delta))$ is a po-$*$-sfield with (SRP). Upon replacement of K with $K((x;\delta))$ this construction can be iterated, and is a rich source of diverse examples.

The well known theorem that the total order on a commutative real closed field is unique is an immediate consequence of the fact that positive elements have square roots. From the latter it could be expected that in a partially ordered division ring with (SRP) some partial order (contained in the given one) might be in some sense unique. If a po-$*$-sfield has (SRP) then every symmetric element $s = s^*$ is of the form $s = \pm d^2$. This latter property ($s = s^* \Longrightarrow s = \pm d^2$) alone could possibly be used to construct a partial order on the division ring. It is beyond the scope of this article to formulate this precisely (see 2.14), but the author plans to return to both questions in the future. However, first it has to be shown that such nontrivial division rings with (SRP) exist, and this is a major objective of the present article.

The ordinary Heisenberg algebra of quantum mechanics is the $\mathbf{C}$-algebra $\mathbf{C}\langle p, q\rangle$ of all noncommutative polynomials in p and q subject to $qp - pq = \sqrt{-1}$. All formal Laurent power series ring $\mathbf{C}\langle\langle p, q\rangle\rangle$ with only a finite number of positive exponent terms subject to the same relation $qp - pq = \sqrt{-1}$ is a division ring containing $\mathbf{C}\langle p, q\rangle \subset \mathbf{C}\langle\langle p, q\rangle\rangle$. The division ring $\mathbf{C}\langle\langle p, q\rangle\rangle$ is a very special and simple example of the above more general class of po-$*$-sfields $K((x;\delta))$.

The square root axiom (SR) requires only that elements of the form $\alpha^*\alpha$ have symmetric square roots $\alpha^*\alpha = \beta^2$ unique up to ± 1 (see 1.5). Clearly, (SRP) implies (SR).

J. Martinez (ed.), Ordered Algebraic Structures, 209–228.

The reals $\mathbf{R}$, complexes $\mathbf{C}$, and standard real quaternions $\mathbf{H_R}$ (and easy modifications of these) are standard examples of po-$*$-sfields with (SR). For all of these, $\alpha < \beta$ if and only if $0 < \beta - \alpha \in \mathbf{R}^+$. There is an acceptable theory of $*$-sfields with (SR) ([H1; p.214]) – but only essentially one single solitary (nonstandard) example of such a division ring, namely $\mathbf{C}\langle\langle p, q\rangle\rangle$ ([H1; p.213-219]).

Needless to say in $\mathbf{C}\langle\langle p, q\rangle\rangle$, $p^* = p$ and $q^* = q$ are self adjoint. Here we also construct classes of po-$*$-sfields $K((x;\delta))$ where $x^* = -x$ is skew. There is a real problem here in going from a commutative scalar field like $\mathbf{C}$ to a noncommutative K. The verification of the (SR) property requires one to solve the equation $a\xi + \xi a = d$ where $a, d \in K$, for an unknown $\xi \in K$. Even in the quaternions $K = \mathbf{H_R}$ this equation cannot be solved for all choices of $a, d \in \mathbf{H_R}$.

To summarize, there are two objectives. I. First, to construct a class of po-$*$-sfields with (SRP) such that the Heisenberg division algebra $\mathbf{C}\langle\langle p, q\rangle\rangle$ is a special case. The latter, and in particular its partial order, has potential in as yet unexplored applications in physics. II. Therefore, perhaps just as important as I is the construction and systematic development of various computational algorithms (valid in the class I and in particular in $\mathbf{C}\langle\langle p, q\rangle\rangle$). For example, for any $\alpha \in K((x;\delta))$ or $\mathbf{C}\langle\langle p, q\rangle\rangle$, these algorithms will compute (up to any required finite number of terms in a power series expansion) the polar decomposition of $\alpha = \sqrt{\alpha^*\alpha}u = u\sqrt{\alpha^*\alpha}$, where u is unitary ($u^*u = uu^* = 1$), and $0 < \sqrt{\alpha^*\alpha}$.

Section 3 continues the study of division ring extensions $K \subset D$ already begun in section 2, where K was extended to $K \subset K((x;\delta))$. In section 3, the involution and the partial order is lifted from K to a twisted group power series extension $K[[\Gamma;\theta]] \supset K$. In the theory of division rings, frequently the number of theorems exceeds by far the known examples. The total number of essentially different concrete examples of po-$*$-sfields in the literature seems to be very small ([Ch 1, 2, 3], [H 1, 2, 3], [I], and [Cr 1, 2, 3]). The objective of section 3 is to create new kinds of hitherto unknown po-$*$-sfields, by construcing general classes of po-$*$-sfields which together with section 2, include as special cases almost all of the examples known so far.

All the examples D po-$*$-sfields here have a natural valuation $v: D\backslash\{0\} \longrightarrow \Gamma$ with $\Gamma \subset D$, and contain the residue skew field $K \subset D$. The author believes that there are structure theorems for partially ordered division rings D of the sort which under appropriate hypotheses embed $K \subset K[\Gamma;\theta] \subset D \subset K[[\Gamma;\theta]]$ (or $K \subset K[x;\delta] \subset D \subset K((x;\delta))$), roughly speaking. The starting point for such an embedding theorem must be a valuation $v: D\backslash\{0\} \longrightarrow \Gamma$ whose residue skew field K is a subsfield of D. For this purpose, the usefulness of valuations frequently seems to be limited if the residue sfield is small, i.e. $\mathbf{R}$ or a subfield of $\mathbf{C}$.

Valuations on division rings have many applications ([Ch 1], [Co 1], [Cr 1], [D2], and [H2]) and give much information about a division ring. Frequently in the literature, some authors start with a division ring D and assume as a hypothesis that there exists a totally ordered group Γ and a valuation $v: D\backslash\{0\} \longrightarrow \Gamma$ whose residue sfield is partially ordered, and then lift this partial order from the residue field back to all of D (e.g. [Cr 1, 2], [H 2], see also [Co 1]). One of the objectives here will be to show that these hypotheses can actually sometimes be satisfied by a partially ordered division ring. Conversely, we start with a partially ordered division ring D, an appropriate skew subfield $K \subset D$, and then manufacture out of $K \subset D$ and from the partial order of D a totally ordered group Γ and a valuation $v: D\backslash\{0\} \longrightarrow \Gamma$ whose residue skew field contains a canonical order isomorphic copy of K.

1. PRELIMINARIES

A $*$-sfield (star-sfield) D is a division ring D with an involution "$*$" satisfying: $a^{**} = a$, $(a-b)^* = a^* - b^*$, and $(ab)^* = b^*a^*$ for all $a, b \in D$. Throughout, P denotes the cone of strictly positive elements of an additive partial order on a division ring D, where $a < b$ if and only if $b - a \in P$. Let $\mathrm{sym}(D)$ denote the symmetric elements $\mathrm{sym}(D) = \{s \in D \mid s^* = s\}$.

1.1. DEFINITION. A po-$*$-sfield is a division ring with an involution $(D, *)$ and a partial order P satisfying the following

(a) $P^* = \{a^* \mid a \in P\} = P$; (b) $P + P \subseteq P$;

(c) $P \cap -P = \emptyset$; (d) $0 < 1 \in P$;

(e) $d^*Pd \subseteq P$ for all $0 \neq d \in D$.

(f) $\mathrm{sym}(D)\backslash\{0\} \subseteq P \cup -P$, that is the set of symmetric elements of D is a totally ordered set.

These are minimal hypotheses which will be satisfied by all partial orders. From now on the notation "$(D, *, P)$" means that D is a division ring satisfying the above six conditions.

1.2. REMARKS. For $(D, *, P)$ satisfying 1.1(a)-(f), set $\mathrm{sym}(D)^+ = \{a \mid 0 < a^* = a\}$, center $D = \{c \in D \mid cd = dc$ for all $d \in D\}$ and let center $D \supset \mathbf{Q}$ be the rationals. If $0 < a = a^* \in P$, then

(1) if $b = b^* \in D$ such that $ab = ba$, then $0 < b^*ab = ab^2$. Hence also $0 < b^2$.

(2) $\mathrm{sym}(D)^{+-1} \subset P$, for by (1), $0 < a(a^{-1})^2 = a^{-1} \in P$.

(3) By 1.1(d) and (2), $\mathbf{Q}^+ \subset P$.

(4) For $p/q \in \mathbf{Q}^+$, $0 < [q(pa)^{-1}]^{-1} = (p/q)a$.

(5) If $0 < a = a^* < 1$, then $1 - a \in \mathrm{sym}(D)^+$, and by (1), $0 < a^2(1-a) + a(1-a)^2$. Hence $0 < a^2 < a$.

(6) If $1 < a = a^*$, then $a^{-1} < 1$, because $0 < (a^{-1})^2(a-1) + [(a-1)a^{-1}]^2 = 1 - a^{-1}$.

1.3. DEFINITIONS. For $(D, *, P)$ satisfying 1.1(a)-(f), P is a Baer ordering if (f*) sym$(D)\backslash\{0\} = P \cup -P$. If D is commutative and $*$ is the identity, then a Baer order reduces to a semiordering in the sense of [P2; p.1 and p.5]. (See also [P1; p.320] and [GP; p.17-18].) If $(D,*,P)$ is a po-$*$-sfield, then $D^+ \equiv P \cap \text{sym}(D)$ gives a Baer order on D.

A Baer ordering is a Jordan ordering if

(j) $\forall\ a,b \in P,\ ab+ba \in P$.

Any P satisfying 1.1(a)-(f) is called a strong ordering if

(g) $PP \subset P$, i.e. if D is a partially ordered ring. (Thus if $0 < a \in P$, then also $a^{-1} = a^*(a^{*-1}a^{-1}) \in P$; and $d^{-1}ad = (d^*d)^{-1}d^*ad \in P$ for any $d \in D$. I.e. $P \triangleleft D\backslash\{0\}$. If $0 < a < b$, then $0 < b^{-1} < a^{-1} \in P$.)

Any po-$*$-sfield $(D,*,P)$ induces a Baer order on D, i.e. $-\ P \cap \text{sym}(D)$. There are two reasons for formulating everything in terms of po-$*$-sfields, rather than Baer orders. The po-$*$-sfields are more general; any Baer ordered division ring automatically is a po-$*$-sfield. But more importantly, we want to use partial orders for which, given any element $\alpha \in D$, it is possible to verify whether $0 < \alpha$ or not. This is the case here in all examples of po-$*$-sfields, and never the case in all nontrivial examples of Baer - ordered division rings used here. Frequently, in the latter case the condition $0 < \alpha$ is equivalent to an infinite number of equations in an infinite number of unknowns (see 2.10).

1.4. DEFINITION. Let $(D,*)$ be a $*$-sfield and $n = 1,2,\ldots; a_1,a_2,\ldots,a_n; x = x* \in D\backslash\{0\}$ arbitrary. Then $(D,*)$ is

weakly formally real *if* $a_1^*a_1 + \ldots + a_n^*a_n \neq 0$; and
formally real *if* $a_1^*xa_1 + \ldots + a_n^*xa_n \neq 0$.

It has been shown in [H3; Theorem 3.9] that an involutive division ring has a strong order if and only if it is formally real. Although the next definition does not require D to be partially ordered, so far all known examples having (SR) can be strongly ordered.

1.5. DEFINITION. A$*$-sfield $(D,*)$ satisfies (SR) if

(SR) $\forall\ \alpha \in D, \exists$ a solution $x = \beta = \beta^*$ of $x^2 - \alpha^*\alpha = 0,\ \beta^2 = \alpha^*\alpha$; and if $\gamma \in D,\ \gamma^2 - \alpha^*\alpha = 0$, then $\Longrightarrow \gamma = \pm\beta$.

1.6. DEFINITION. A po-$*$-sfield $(D,*,P)$ with 1.1(a)-(f) satisfies (SRP) if

(SRP) $\forall\ 0 < \alpha = \alpha^* \in D,\ \exists\beta = \beta^*,\ \beta^2 = \alpha$; and any solution $x \in D$ of $x^2 = \alpha$ is $x = \pm\beta$. (Hence by 1.1(f), either $0 < \beta$ or $\beta < 0$.) Clearly, (SRP) $\Longrightarrow$ (SR).

1.7. DEFINITION. A po-$*$-sfield K satisfies the linear equations property (LE) if

(LE) $\forall\ d_0 \in K$, $\forall\ 0 < a_0 = a_0^* \in K$, the equation $\xi a_0 + a_0 \xi = d_0$ has a *unique* solution $\xi = b \in K$, i.e. $ba_0 + a_0 b = d_0$.

The method of proof in [H1; p.213-214] can be used to show the following.

1.8. REMARK. Let D satisfy (SRP). Then the above β commutes with every element of D that commutes with α. In particular, $\beta\alpha = \alpha\beta$ and $\beta\alpha^{-1} = \alpha^{-1}\beta$.

In concrete examples, to verify that $\beta^* = \beta$ may be difficult, and the next lemma does this indirectly with the help of the partial order.

1.9. LEMMA. Assume that $(D, *, P)$ satisfies 1.1(a)-(f), and that for every $0 < \alpha = \alpha^* \in D$, there exists a $\beta \in D$ unique up to ± 1 with $\beta^2 = \alpha$. Then D satisfies (SRP).

PROOF. Since $\beta^{*2} = \alpha^* = \alpha$, by uniqueness, $\beta^* = \pm\beta$. If $\beta^* = -\beta$, then $0 = \beta^*\beta + \alpha \in P + P \subseteq P$, a contradiction. Hence $\beta^* = \beta$.

2. SQUARE ROOTS OF POSITIVE ELEMENTS

Suppose that K is a po-$*$-sfield, and $K \subset K((x;\delta))$ a skew power series division ring extension of K. First it is shown how the involution and partial order can be extended from K to $K((x;\delta))$.

2.1. SKEW POWER SERIES DIVISION RINGS. Let K be a sfield, $\delta: K \longrightarrow K$ a derivation $((ab)^\delta = a^\delta b + ab^\delta,\ a, b \in K)$ and set $\delta(i) = \delta^i$ for $i \in \mathbf{Z}^+ \cup \{0\} = 0, 1, 2, \ldots$; where $\delta(0) =$ identity. Denote all the formal Laurent power series $\alpha = \sum\{x^i a_i \mid i \leq p\}$, $a_i \in K$, $p \in \mathbf{Z}$ by $K((x;\delta))$. Set $\operatorname{supp}\alpha = \{x^i \mid a_i \neq 0\}$. Then this is known to be a division ring ([D1, p.203]) under formal power series term by term multiplication subject to the relation $kx = xk + k^\delta$. Throughout, define $y = x^{-1}$. Consequently $ky = yk - yk^\delta y = yk - y^2 k^\delta + y^2 k^{\delta\delta} y$, etc.

Every element α of $K((x;\delta))$ is of the form $\alpha = x^p \sum\{x^{-i} c_i \mid i = 0, 1, 2, \ldots\} = x^p(1 - \lambda)c_0$ where $c_i \in K$, $c_0 \neq 0$, $p \in \mathbf{Z} = 0, \pm 1, \pm 2, \ldots$; and $\lambda = -\sum_1^\infty y^i c_i c_0^{-1}$. Thus $\alpha^{-1} = 1/\alpha = c_0^{-1}(1 - \lambda)^{-1} x^{-p}$, where $(1 - \lambda)^{-1} = 1 + \lambda + \lambda^2 + \ldots$ is a formal power series in y with nonegative exponents only.

If $\delta = 0$, then write $K((x;0)) = K((x))$ in which case $kx = xk$ for all $k \in K$. Recall that for any α (in $\mathbf{Z}$, K, or $K((x;\delta))$) and any $k = 0, 1, 2, \ldots$; $\binom{\alpha}{k} = [\alpha(\alpha - 1) \ldots (\alpha - k + 1)]/k!$, $\binom{\alpha}{0} = 1$, $\binom{0}{0} = 1$.

The next two lemmas can be proved easily by induction, and throughout we prove only left to right identities.

2.2. LEMMA. For $c \in K$ and $y = x^{-1} \in D = K((x;\delta))$,

(a) $cy = y\sum_{i=0}^{n-1} y^i c^{\delta(i)}(-1)^i + y^n c^{\delta(n)}(-1)^n y = y\sum_0^\infty y^i c^{\delta(i)}(-1)^i$

(b) $yc = \sum_{i=0}^{n-1} c^{\delta(i)} y^i y + yc^{\delta(n)} y^{n-1} = \sum_{i=0}^\infty c^{\delta(i)} y^i y.$

2.3. LEMMA. For any $m = 1, 2, \ldots$; for any $k = 0, 1, 2, \ldots$

$$\sum_{i=0}^{k} \frac{(m-1+i)!}{(m-1)!i!} = \frac{(m+k)!}{m!k!} = (-1)^k \binom{-m-1}{k}.$$

2.4. PROPOSITION. For any $c \in K$ and any $m = 0, 1, 2, \ldots$ the following hold.

(a) $cx^m = x^m \sum_{k=0}^m x^{-k} \binom{m}{k} c^{\delta(k)}$;

(b) $x^m c = \sum_{k=0}^m (-1)^k c^{\delta(k)} \binom{m}{k} x^{-k} x^m$;

(c) $cy^m = y^m \sum_{k=0}^\infty y^k \binom{-m}{k} c^{\delta(k)}$

(d) $y^m c = \sum_{k=0}^\infty c^{\delta(k)} (-1)^k \binom{-m}{k} y^k y^m$.

PROOF. (a) Use induction, change of summation variable, and $\binom{m-1}{j} + \binom{m-1}{j-1} = \binom{m}{j}$. (c) By induction, assume that (c) holds for m, and use 2.3:

$$\begin{aligned}
(cy^m)y &= [y^m \sum_{i=0}^{\infty} y^i \frac{(m-1+i)!}{(m-1)!i!}(-1)^i c^{\delta(i)}]y \\
&= y^m \sum_{i=0}^{\infty} y^i \frac{(m-1+i)!}{(m-1)!i!}(-1)^i y \sum_{j=0}^{\infty} y^j c^{\delta(i+j)}(-1)^j \\
&= y^{m+1} \sum_{k=0}^{\infty} y^k (-1)^k c^{\delta(k)} \sum_{(i,j) i+j=k} \frac{(m-1+i)!}{(m-1)!i!} \\
&= y^{m+1} \sum_{k=0}^{\infty} y^k c^{\delta(k)} (-1)^k \sum_{i=0}^{k} \frac{(m-1+i)!}{(m-1)!i!} \\
&= y^{m+1} \sum_{k=0}^{\infty} y^k c^{\delta(k)} \binom{-m-1}{k}.
\end{aligned}$$

2.5. MAIN COROLLARY. For any $c \in K$ and any $m = 0, \pm 1, \pm 2, \ldots$

$$cx^m = x^m \sum_{k=0}^{\infty} x^{-k} \binom{m}{k} c^{\delta(k)}.$$

$$x^m c = \sum_{k=0}^{\infty}(-1)^k c^{\delta(k)} \binom{m}{k} x^{-k} x^m.$$

Our next objective is to start with a po-∗-sfield K with a derivation δ, and then extend (i) the involution, (ii) the derivation and (iii) partial order to all of $K((x;\delta))$.

2.6. DEFINITION. For a ∗-sfield K and a derivation $\delta: K \longrightarrow K$, δ is

$$\underline{\text{symmetric}}: \text{ if } \delta* = *\delta,\ k^{*\delta} = k^{\delta*}; \text{ and}$$
$$\underline{\text{skew}}: \text{ if } \delta* = -*\delta,\ k^{*\delta} = -k^{\delta*}, k \in K.$$

For $s \in$ center K, let $s\delta: K \longrightarrow K$ be the derivation of K defined by $k \longrightarrow s(k^\delta)$. Then

$s* = s \in \text{center}K$, δ symmetric $\Longrightarrow s\delta$ is symmetric;
$s* = -s \in \text{center}K$, δ skew $\Longrightarrow s\delta$ is symmetric; and
$s* = -s \in \text{center}K$, δ symmetric $\Longrightarrow s\delta$ is skew.

2.7. THEOREM. For a ∗-sfield K, let $\delta: K \longrightarrow K$ be either a skew or a symmetric derivation. Let $\alpha = \sum\{x^i a_i \mid -\infty < i \leq N\} \in K((x;\delta))$, $0 \neq a_N \in K$. Define functions $\partial, *: K((x;\delta)) \longrightarrow K((x;\delta))$ by

$$\partial\alpha = \alpha^\partial = \sum_i x^i a_i^\delta;$$

(a) if δ is skew, $x^* = x$, $\alpha^* = \sum_i a_i^* x^i$;
(b) if δ is symmetric, $x^* = -x$, $\alpha^* = \sum a_i^*(-1)^i x^i$.

If $\delta = 0$, define "∗" by either (a) or (b). Then

(i) "∗" is an extension of the involution from K to $K((x;\delta))$.
(ii) ∂ is a derivation extending δ from K to $K((x;\delta))$. If δ is skew (or symmetric) so is also ∂.

Now in addition assume that $(K, *, P)$ satisfies 1.1 (a)-(f), and define $0 < \alpha$ if $0 < a_N \in K$. Then $K((x;\delta))$ has property $(\mathcal{P})$ if K has property $(\mathcal{P})$ where $(\mathcal{P})$ can be any one of the following:

(iii) po-∗-sfield, strong ordering; or formally real. A Baer ordering can be extended from K to $K((x;\delta))$ by defining $0 < \alpha$ if in addition to $0 < a_N \in K$, also $\alpha^* = \alpha$.

PROOF. The proof requires repeated use of 2.5 and is omitted.

2.8. From now on $D = K((x;\delta))$ and the involution "∗" on D are as in the last theorem, where δ is either skew, or symmetric. Unless indicated otherwise, in sums, i, j, k, n range over $0, 1, 2, \ldots$ only.

One next objective is to show that $K((x;\delta))$ inherits (SR) and (SRP) from K. Perhaps just as important as these are the computational algorithms developed along the way. The hypothesis below is satisfied by every commutative field K of characteristic $\neq 2$.

2.9. PROPOSITION. Assume that K is a skew field and $0 \neq a_0 \in K$ has the property that for any $d_0 \in K$, the equation $ba_0 + a_0 b = d_0$ has a solution $b \in K$. Let $\alpha = x^p \sum_0^\infty x^{-i} a_i$, $\gamma = x^s \sum_0^\infty x^{-i} d_i \in K((x;\delta))$ where $a_1, a_2, \ldots; d_0, d_1, \ldots \in K$ are arbitrary. Then the equation
(i) $\beta\alpha + \alpha\beta = \gamma$ has a solution $\beta \in K((x;\delta))$.
(ii) $b \in K$ unique $\Longrightarrow \beta \in K((x;\delta))$ unique.

PROOF. First note that every $\alpha \neq 0$ is necessarily of the above form with $a_0 \neq 0$ for some unique p. We solve $\beta\alpha + \alpha\beta = \gamma$ for $\beta = x^q \sum_0^\infty x^{-i} b_i$. First, $s = p + q$. By 2.4 and a summation variable change, we get from

$$\beta\alpha + \beta\alpha = x^q \sum_0^\infty x^{-i} b_i \sum_0^\infty x^{p-j} a_j + x^p \sum_0^\infty x^{-j} a_j \sum_0^\infty x^{q-i} b_i \text{ that}$$

$$\beta\alpha + \alpha\beta = x^{p+q} \sum_{n=0}^\infty x^{-n} \{ \sum_{\substack{(i,j,k) \\ i+j+k=n}} \binom{p-j}{k} b_i^{\delta(k)} a_j + \binom{q-i}{k} a_j^{\delta(k)} b_i \}$$

$$= x^{p+q} \sum_{n=0}^\infty x^{-n} \{ \sum_{\substack{(i,j) \\ i+j=n}} (b_i a_j + a_j b_i) + S_n \}, \text{ where}$$

$$S_n = \sum_{k=1}^n \sum_{\substack{(i,j) \\ i+j=n-k}} \binom{p-j}{k} b_i^{\delta(k)} a_j + \binom{q-i}{k} a_j^{\delta(k)} b_i.$$

There is no b_n in S_n, only $b_0, b_1, \ldots, b_{n-1}$ and their derivatives. Next

$$\beta\alpha + \alpha\beta = x^{p+q} \sum_{n=0}^\infty x^{-n} \{ b_n a_0 + a_0 b_n + T_n + S_n \},$$

$$T_n = \sum_{\substack{(i,j) \\ 0 \leq i \leq n-1, i+j=n}} b_i a_j + a_j b_i.$$

Again, there are no b_n's in T_n. Note that

$$\beta\alpha + \alpha\beta = x^{p+q} \{ [b_0 a_0 + a_0 b_0] + \\ + x^{-1} [b_0 a_1 + a_1 b_0 + b_1 a_0 + a_0 b_1 + (b_0^\delta a_0 + a_0^\delta b_0)] + x^{-2} [(b_2 a_0 + a_0 b_2) \\ + (b_1 a_1 + a_1 b_1 + b_0 a_2 + a_2 b_0 + b_1^\delta a_0 + a_0^\delta b_1 + b_0^\delta a_1 + a_1^\delta b_0)] + \ldots \}$$

Set $\beta\alpha+\alpha\beta=\gamma$. By hypotheses, we can find $b_0\in K$ such that $b_0a_0+a_0b_0=d_0$. Next, solve for $b_1\in K$ the equation $b_1a_0+a_0b_1=d_1-(b_0^\delta a_0+a_0^\delta b_0)$. An induction now proves (i), and uniqueness in K implies (ii).

The next computation shows that the right side K-coefficients of the positive elements $\alpha^*\alpha$ are not even symmetric in general, except for the top coefficient.

2.10. Let $\alpha=x^p\sum_0^\infty x^{-i}a_i\in D$ and let δ be skew. Then by 4.3 (c) and a change of summation variable

$$\begin{aligned}\alpha^*\alpha&=x^{2p}\sum_{n=0}^{\infty}x^{-n}\sum_{\substack{(i,j,k)\\ i+j+k=n,0\le i,j,k\le n}}\binom{-i-j+2p}{k}a_i^{*\delta(k)}a_j\\ &=x^{2p}\{a_0^*a_0+x^{-1}2p(a_1^*a_0+a_0^*a_1+a_1^*\delta a_0)+\ldots\}.\end{aligned}$$

2.11. THEOREM. Let $(K,*,P)$ be a po-$*$-sfield, $\delta\colon K\longrightarrow K$ a skew or symmetric derivation, and $(K((x;\delta)),*)$ as in 2.7. If K satisfies either the (SR)+(LE), or (SRP)+(LE) axioms, then $K((x;\delta))$ does likewise.

PROOF. We prove more. Let $0\ne\gamma=x^{2p}\sum_0^\infty x^{-i}a_i\in D$ be any element such that $0\ne a_0=b_0^2$, $b_0^*=b_0\in K$ for some $b_0\in K$ unique up to ±1. Assume that the equation $bb_0+b_0b=d_0$ has a unique solution $b\in K$ for every choice of d_0.

We let $\beta=x^p\sum_0^\infty x^{-i}b_i$ and (i) solve $\beta^2=\gamma$ for the b_i in terms of the a_i, and show that $\pm\beta$ are the only such solutions. (ii) We show that replacement of b_0 with $-b_0$ replaces β with $-\beta$. (iii) Lastly, it is shown that under the additional hypothesis that $\gamma=\gamma^*$, also $\beta^*=\beta$.

Use of 2.4(a) and (c) followed by a summation variable change and some reshuffling gives

$$\begin{aligned}\beta^2&=\sum_{(i,j)}x^{-i+p}[b_ix^{-j+p}]b_j=x^{2p}\sum_{(i,j)}\sum_{k=0}^{\infty}x^{-i-j-k}\binom{-j+p}{k}b_i^{\delta(k)}b_j\\ &=x^{2p}\sum_{n=0}^{\infty}x^{-n}\sum_{\substack{(i,j,k)\\ i+j+k=n}}\binom{-j+p}{k}b_i^{\delta(k)}b_j\\ &=x^{2p}b_0b_0+x^{2p}\sum_{n=1}^{\infty}x^{-n}\{\sum_{\substack{(i,j)\\ i+j=n}}b_ib_j+\sum_{k=1}^{n}\sum_{\substack{(i,j)\\ i+j=n-k}}\binom{-j+p}{k}b_i^{\delta(k)}b_j\}\\ &=x^{2p}b_0b_0+\\ &\quad+x^{2p}\sum_{n=1}^{\infty}x^{-n}\{b_nb_0+b_0b_n+\sum_{\substack{(i,j)\\ i+j=n}}b_ib_j+\sum_{k=1}^{n}\sum_{\substack{0\le i,j\le n-1\\ i+j=n-k}}\binom{-j+p}{k}b_i^{\delta(k)}b_j\}\\ &=x^{2p}\{b_0^2+x^{-1}(b_1b_0+b_0b_1+pb_0^\delta b_0)+x^{-2}(\ldots)+\ldots\}.\end{aligned}$$

In order that $\beta^2 = \gamma$, solve $b_1 b_0 + b_0 b_1 = a_1 - p b_0^\delta b_0$ for a unique $b_1 \in K$. Replacement of b_0 in β with $-b_0$ results in $-b_1$.

Assume by induction that $b_0, b_1, \ldots, b_{n-1}$ have been solved for (in terms of $a_0, a_1, \ldots, a_{n-1}$) and are odd in b_0 (replacement of b_0 with $-b_0$ in β gives $-b_0, -b_1, \ldots, -b_{n-1}$). Equate K-coefficients of x^{-n} in $\beta^2 = \gamma$,

$$b_n b_0 + b_0 b_n = a_n - \sum_{\substack{1 \le i,j \le n-1 \\ i+j=n}} b_i b_j - \sum_{k=1}^{n} \sum_{\substack{0 \le i,j \le n-1 \\ i+j=n-k}} \binom{-j+p}{k} b_i^{\delta(k)} b_j.$$

The right side involves no b_n and is an even function of $b_0, b_1, \ldots, b_{n-1}$ and their derivatives. The rest is clear.

The next construction can be iterated any finite number of times, and ω-times, by taking a union.

2.12. COROLLARY. Let K be a po-$*$-sfield (1.1(a)-(f)) with $\delta: K \longrightarrow K$ either a skew, or a symmetric derivation. Set $D = K((x;\delta))$ and let $*: D \longrightarrow D$, $\partial: D \longrightarrow D$, and the partial order on D be as in Theorem 2.7. Let $(\mathcal{P})$ be any one of the following properties: po-$*$-sfield, strong ordering, Baer ordering, Jordan ordering, (SR)+(LE), or (SRP)+(LE), or formally real. Then D and $D((x_2;\partial))$ also have property $(\mathcal{P})$.

2.13. EXAMPLE. For $k(q) = \sum q^j c_j \in \mathbf{C}((q))$, $k(q)^* = \sum q^j \overline{c}_j$ where $\overline{c}_j \in \mathbf{C}$ is the complex conjugate; $\delta k(q) = \sqrt{-1}\frac{d}{dq}k(q)$ formally and $*\delta = -\delta*$. Thus $\mathbf{C}((q))((p;\delta)) = \mathbf{C}\langle\langle p,q\rangle\rangle$ is the Heisenberg division algebra with $k(q)p = pk(q) + \delta k(q)$. Note that $q^* = q$, $p^* = p$, but $(pq)^* = qp = pq + \sqrt{-1} \neq pq$. It is an immediate corollary of 2.7 and 2.11 that $\mathbf{C}\langle\langle p,q\rangle\rangle$ is strongly ordered, and satisfies both (SR) and (SRP).

The above facts about $\mathbf{C}\langle\langle p,q\rangle\rangle$ were proved by S. Holland. The next example cannot be constructed from Holland's arguments (in [H1; p.213-219]) and 2.11, 2.12, and 2.7 must be used for this.

2.14. GENERALIZED HEISENBERG DIVISION RINGS. Let the spatial variables x_1, x_2, x_3 and the associated momentum variables satisfy the usual commutation rules for all $i, j = 1, 2$, and 3:

$$[x_i, p_j] = \sqrt{-1}\delta_{ij},\ [x_i, x_j] = 0,\ [p_i, p_j] = 0.$$

Let $\mathbf{C}((x_1, x_2, x_3))$ be the commutative formal Laurent power series field with only a finite number of positive exponent terms in x_1, x_2, and x_3, and with termwise conjugation of the complex coefficients. Set $\delta_i = \sqrt{-1}\partial/partialx_i$ for $i = 1, 2$, and 3. By iteration of the involution and lexicographic order as in 2.7 (a), $D \equiv \mathbf{C}\langle\langle x_1, x_2, x_3; p_1, p_2, p_3\rangle\rangle = \mathbf{C}((x_1, x_2, x_3))((p_1;\delta_1))((p_2;\delta_2))((p_3;\delta_3))$ is a strongly ordered formally real *-division ring lexicographically ordered according to $x_1 < x_2 < x_3 < p_1 < p_2 < p_3$ with (SRP). The division ring D does not depend on the order in which the iteration is carried out although the partial order does, e.g. $D = \mathbf{C}((p_1, p_2, p_3))((x_1; -\delta_1))((x_2; -\delta_2))((x_3; -\delta_3))$ is the same division ring but with a different lexicographic order in which $p_1 < p_2 < p_3 < x_1 < x_2 < x_3$.

Positive cones on a division ring D yielding strong orders with (SRP) are closed under arbitrary intersections. Thus by intersecting an appropriate finite number of such positive cones on D, we can obtain a partial order on D which treats the x_1, x_2, x_3 and $p_1, p_2,$ and p_3 on an equal footing and still is strongly ordered with (SRP). From a physics point of view, the latter is a minimum requirement for the partial order to have a physical significance.

All of the usual quantum mechanical operators and identities needed to describe the hydrogen atom, or a quantum mechnical harmonic oscillator are contained in D. We list only a few, e.g. if (i, j, k) is a cyclic permutation of $(1, 2, 3)$, then we have

$$L_i \equiv x_j p_k - x_k p_j, \quad [L_i, p_j] = \sqrt{-1} p_k = [p_i, L_j],$$
$$[x_i, L_j] = x_k = [L_i, x_j], \quad [L_i, L_j] = \sqrt{-1} L_k.$$
$$H \equiv \frac{1}{2m}(p_1^2 + p_2^2 + p_3^2) + V \cdot 1, \quad m \in \mathbf{R}; \quad V \in \mathbf{R}((x_1, x_2, x_3)) \subset D,$$
$$\sqrt{-1}[H, p_1] = \sqrt{-1}[V, p_1] = \sqrt{-1}\delta_1 V \equiv \frac{d}{dt} p_1 \in \mathbf{R}((x_1, x_2, x_3)) \subset D, \ t = \text{time}.$$

Above, clearly "3" can be replaced by any finite integer n (or $n = \omega$).

Some algebras of importance in physics either contain or are built out of quaternion algebras (see [D3]) and the involutions sometimes have a physical significance. The next division algebra is new and is a natural extension of the Heisenberg division algebra.

2.15. EXAMPLE. In $\mathbf{H_R} = \mathbf{R} + \mathbf{RI} + \mathbf{RJ} + \mathbf{RIJ}$ the equation $\xi a + a\xi = d$ can be solved for ξ for every choice of $d \in \mathbf{H_R}$ provided that $a^* \neq -a$. Also, $\mathbf{H_R}$ satisfies (SRP) and hence (SR). For $h = r_0 + r_1 I + r_2 J + r_3 IJ \in \mathbf{H_R}$ with $r_i \in \mathbf{R}$, $h^* = r_0 - r_1 I - r_2 J - r_3 IJ$. View $\mathbf{C} \subset \mathbf{H_R}$ with $\mathbf{C} = \mathbf{R} + \mathbf{RI}$ where $I = \sqrt{-1}$. For $k(q) = \sum\{q^j a_j \mid -\infty < j \leq N\} \in \mathbf{H_R}((q))$ where $a_j \in \mathbf{H_R}$, $a_N \neq 0$, $N \in \mathbf{Z} = 0, \pm 1, \pm 2$; define $k(q)^* = \sum q^j a_j^*$ and $\delta k(q) = I \frac{d}{dq} k(q)$. Set $D = \mathbf{H_R}((q))((p; \delta))$. Then $\mathbf{C}\langle\langle p, q\rangle\rangle \subset D$.

Let $\alpha = \sum\{p^i k_i \mid -\infty < i \leq M\}$ where $k_i \in \mathbf{H_R}((q))$, $k_M = k(q)$ exactly as above, and $M \in \mathbf{Z}$. Thus $\alpha^* = \sum k_i^* p^i$. Define

$$0 < h \Longleftrightarrow 0 < h = r_0 \in \mathbf{R}; \ 0 < k(q) \Longleftrightarrow 0 < a_N \in \mathbf{R};$$
$$0 < \alpha \Longleftrightarrow 0 < k(q) \in \mathbf{H_R}((q))^+.$$

This is a strong ordering on D (and hence D is a po-ring with $0 < \alpha, \beta \Longrightarrow 0 < \alpha\beta$); D satisfies (SR) and (SRP).

As in 2.11, we can also impose a Baer ordering on D which is also a Jordan order; D is formally real in all orderings.

2.16. EXAMPLE. For $k(t) = \sum\{t^j c_j \mid -\infty < j \leq M\} \in \mathbf{C}((t))$, as before in 2.13, $k(t)^* = \sum t^j \bar{c}_j$, $t^* = t$, but now $\delta k(t) = \frac{d}{dt}k(t)$. Set $D = \mathbf{C}((t))((x;\delta))$ where $k(t)x - xk(t) = \delta k(t)$. Now $\delta * = *\delta$ and hence $x^* = -x$. Note that $\mathbf{R}((x^2)) \subset \mathrm{sym}(D)$, but also $(xt\sqrt{-1} + \sqrt{-1}/2)^* = xt\sqrt{-1} + \sqrt{-1}/2$. As in the last example, this D has every property $(\mathcal{P})$ listed in 2.12. Although the above essentially Weyl division algebra is well known ([Ch1; p. 520]), to the authors best knowledge, the fact that it satisfies (SRP), as shown here is new.

3. PO-*-SFIELD EXTENSIONS

It is shown how the partial order and involution can be extended from K to a twisted group power series division ring $K[[\Gamma;\theta]]$ over K.

3.1. NOTATION. For a skew field and a totally ordered group Γ, suppose that $\theta: \Gamma \longrightarrow \mathrm{Aut}\, K$ is a group homomorphism into the automorphism group $\mathrm{Aut}\, K$ of K. For $x, y \in \Gamma$ and any $k, c \in K$, write:

$$x\theta: K \longrightarrow K,\ (xy)\theta = (x\theta)(y\theta),\ e = 1 \in \Gamma;$$
$$k^x \equiv k(x\theta) = k^{x\theta},\ k((xy)\theta) = (k^x)^y,\ k^e = k;$$
$$(k-c)^x = k^x - c^x,\ (kc)^x = k^x c^x;$$
$$k^{-x} \equiv 1/k^x,\ k^{-x} = \left(\frac{1}{k}\right)^x = (k^{-1})^x = (k^x)^{-1}.$$

Define $K[[\Gamma;\theta]]$ as the set of all formal sums $\alpha = \sum\{x\alpha(x) \mid x \in \Gamma\} = \sum x\alpha(x)$, $\beta = \sum y\beta(y)$; $\alpha(x), \beta(y) \in K$; whose supports $\mathrm{supp}\,\alpha = \{x \in \Gamma \mid 0 \neq \alpha(x) \in K\}$ satisfy the ascending chain condition (A.C.C.). Then $K[[\Gamma;\theta]]$ is called the twisted group power series division ring, where addition is pointwise and

$$xayb = xya^y b,\ ay = ya^y,\ ya = a^{1/y}y;\ a, b \in K$$
$$\alpha\beta = \sum_{(x,y)\in\Gamma\times\Gamma} x\alpha(x)y\beta(y) = \sum_{x,y} xy\alpha(x)^y\beta(y) = \sum_{z\in\Gamma} z \sum_{(x,y),z=xy} \alpha(x)^y(y).$$

Now let $x = \max \mathrm{supp}\alpha$ and $a = \alpha(x)$. Thus $\alpha = xa + \ell.o.t.$, where $\ell.o.t. = \alpha - xa$ stands for lower order terms. The function $w: K[[\Gamma;\theta]] \longrightarrow \Gamma$, $w\alpha = x$ if $\alpha \neq 0$, is a valuation, and

$$\alpha = xa(1-\lambda),\ \lambda = -\sum_{y<e} a^{-1}y\alpha(xy) = -\sum_{y<e} ya^{-y}\alpha(xy),$$
$$\alpha^{-1} = (1-\lambda)^{-1}a^{-1}x^{-1},\ (1-\lambda)^{-1} = 1 + \lambda + \lambda^2 + \ldots.$$

The twisted group ring $K[\Gamma;\theta]$ consisting of all elements whose support is finite is a subring of $K[[\Gamma;\theta]]$. If θ is the identity, write $K[\Gamma] \subset K[[\Gamma]]$, where $kx = xk$ for all $k \in K$.

If Γ is cyclic $\Gamma = \{\dots, x^{-1}, e, x, x^2, \dots\}$ define $x\theta = \theta$, $(x\theta)^n = (x^n)\theta = \theta(n)$ for $n = 0, \pm 1, \pm 2, \dots$. Thus if $m = 0, 1, 2,$ then

$$kx = xk^{\theta},\ kx^{-1} = x^{-1}k^{\theta(-1)},\ k^{-\theta} = (1/k)^{\theta},\ k \in K;$$
$$kx^m = xk^{\theta(m)},\ kx^{-m} = x^{-m}k^{\theta(-m)}.$$

Write $K[\Gamma,\theta] = K(x;\theta)$; and $K[[\Gamma;\theta]] = K((x;\theta))$ is called the twisted Laurent power series ring.

3.2. Suppose that $H \triangleleft \Gamma$ is a normal subgroup with either $|\Gamma/H| = 2$ or $H = \Gamma$. For $x, y \in \Gamma$, define $(-1)^x = -1$ if $xH \neq H$, and $(-1)^x = 1$ if $xH = H$. Then $(-1)^{xy} = (-1)^x(-1)^y$, and $\Gamma/H \longrightarrow \{1, -1\}$, $xH \longrightarrow (-1)^x$ is a group homomorphism.

3.3. THEOREM. Let $D = K[[\Gamma;\theta]]$ be a twisted group power series division ring over a skew field $(K, *)$ with an involution. Suppose that

(a) $H \triangleleft \Gamma$, $|\Gamma/H| = 2$ or $H = \Gamma$; and there is an involution

(b) ${}^{\circ}\colon \Gamma \longrightarrow \Gamma$, $x \longrightarrow x^0$, $(xy)^0 = y^0x^0$, $x^{00} = x$ for $x, y, \in \Gamma$, such that

(c) $k^{y*y^0} = k^*$ for all $k \in K, y \in \Gamma$. Define a map $\#D \longrightarrow D$ by

(d) $\beta = \sum_y yb_y$, $\beta^\# = \sum(-1)^y y^0 b_y^{*y^0}$, $y^\# = (-1)^y y^0$, $b_y \in K$.

Then

(i) $\#$ is an involution which extends $*$ from K to D.

(ii) Conversely, if $\#\colon D \longrightarrow D$ is an involution such that $k^\# = k^*$ for $k \in K$, and $\Gamma^\# \subset \Gamma \cup -\Gamma$, then $\#$ is necessarily of the form (a), (b), (c), and (d).

PROOF. (i) It suffices to show that $[xayb]^\# = (yb)^\#(xa)^\#$.

First,

$$(yb)^\#(xa)^\# \equiv [(-1)^y y^0 b^{*y^0}][(-1)^x x^0 a^{*x^0}] = (-1)^{xy} y^0 x^0 b^{*y^0x^0} a^{*x^0}.$$

Second,

$$\begin{aligned}[xayb]^\# &= [xya^y b]^\# \equiv (-1)^{xy}(xy)^0[a^y b]^{*(xy)^0} \\ &= (-1)^{xy} y^0 x^0 [b^* a^{y*}]^{y^0x^0} = (-1)^{xy} y^0 x^0 b^{*y^0x^0}(a^{y*y^0})^{x^0}.\end{aligned}$$

By (c), the two expressions are equal.

(ii) Set $H = \{z \in \Gamma \mid z^\# \in \Gamma\}$. For any $x, y \in \Gamma$, $x^{-1}y \in H$, and $y = xx^{-1}y$. Thus $|\Gamma/H| \leq 2$. Define ${}^{\circ}\colon \Gamma \longrightarrow \Gamma$ by $x^o = (-1)^x x^\#$. Thus $(xy)^o = (-1)^{xy}(xy)^\# = (-1)^y y^\#(-1)^x x^\# = y^0x^0$.

In extending the partial order from a po-$*$-sfield $(K,*,P)$ to $(K[[\Gamma;\theta]],\#)$ the main obstacle is the axiom 1.1(e).

3.4. CONSTRUCTION. Take $(K,*)$ commutative with a Baer ordering, i.e. satisfying 1.1(a)-(e), and 1.3 (f*). Since K is commutative the set $R = \text{sym}(K)$ is a field, by 1.2 (2). Take Γ to be any free abelian group, and totally order it. Let $H = \langle x^2 \mid x \in \Gamma\rangle \equiv \Gamma^2 \neq \Gamma$ be the (free) subgroup generated by squares of elements of Γ. Similarly, $H^2 = \langle h^2 \mid h \in H\rangle$.

The reason for the restrictive hypotheses that K be commutative and Γ free is that any free group can act on K via the involution. More specifically, for any $k \in K$, $y \in \Gamma$, and $h \in H$, take $y^0 = y$ in 3.3, and define

$$k^y = \begin{cases} k^* & y \notin H \\ k & y \in H \end{cases} \quad (-1)^y = \begin{cases} -1 & y \notin H \\ 1 & y \in H \end{cases} \quad \epsilon(h) = \begin{cases} -1 & h \notin H^2 \\ 1 & h \in H^2. \end{cases}$$

The last two functions are (multiplicative) group characters. Since $k^{*y} = k^{y*}$ and $k^h = k$, 3.3(a), (b), and (c) hold. Set $D = K[[\Gamma;*]]$ where $y^\# = (-1)^y y$ and $(yk)^\# = (-1)^y yk^{*y}$. Thus center $D = \text{sym}(D) = R[[H]]$.

For $\gamma = \sum g r_g \in R[[H]]$, set $h = \max \text{supp}\gamma = w(\gamma)$, and define

$$0 < \gamma \iff 0 < \epsilon(h) r_h \in R.$$

Note that if $y \notin H$, that then $0 < y^\# y = -y^2$.

To verify 1.1 (e), let $k \in K$, $y \in \Gamma$, $h \in H$, and $0 < \epsilon(h) r \in R^+$ be arbitrary. Then

$$(yk)^\# hryk = (-1)^y yk^{*y} hryk = yhy(-1)^y k^{*yhy} r^y k;$$
$$r^y = r \in \text{sym}(K),\ k^{*yhy} = k^*,\ 0 < k^* rk \in R.$$

We only need that $\epsilon(yhy) = \epsilon(y^2)\epsilon(h)$ and that $\epsilon(y^2)(-1)^y > 0$ for all $y \in \Gamma$. Thus

$$\epsilon(yhy)(-1)^y k^* rk = \epsilon(h) k^* rk > 0.$$

The rest of 1.1 clearly holds, and $(D,\#)$ is Baer ordered. Furthermore, the order is compatible with the natural valuation $w\colon D\backslash\{0\} \longrightarrow \Gamma$.

If $(K,*)$ is such that R is a totally ordered field $(R^+R^+ \subseteq R^+)$, then our $(D,\#)$ has a strong order which is also a Jordan order. This is the case if $K = \mathbf{C}$.

There exist (actually Pythagorean) subfields K of the reals, and a total linear order on K as a set, such that with the identity involution $K = R$ is Baer ordered (1.1(a)-(f), and 1.3(f*)), but with $K^+K^+ \not\subseteq K^+$ ([GP; 23]). Then $K \subset K[[H]] =$ center $K[[\Gamma;*]]$, where $K[[\Gamma;*]]$ is neither strongly nor Jordan ordered.

The following Baer ordered division ring is new. It contains as a subskewfield an example of [Cr 1; 5.1].

3.5. APPLICATION. Let Γ be the free abelian group on two generators $\Gamma = \langle x, y\rangle$ ordered lexicographically, where $x^i y^j \leq x^p y^q$ if either $i < p$; or $i = p$ but $j \leq q$. Apply 3.4 to $K = \mathbf{C}$ to get a Baer ordered division ring $\mathbf{C}[[\Gamma; *]] = \mathbf{C}((x, y; *))$ whose elements α are formal Laurent series $\alpha = \sum x^i y^j c_{ij}$. For $n \in \mathbf{Z} = 0, \pm 1, \pm 2, \ldots$ define $c^{*[n]}$ to be c conjugated n times. Then

$$cx = x\overline{c},\ cy = y\overline{c},\ cx^{-1} = x^{-1}\overline{c},\ cy^{-1} = y^{-1}\overline{c},\ xy = yx,$$
$$x^* = -x,\ y^* = -y,\ \alpha^* = \sum (-1)^{i+j} x^i y^j c_{ij}^{*[1+i+j]};$$
$$(xc)^* = -xc,\ (x^2c)^* = x^2\overline{c},\ (xyc)^* = xy\overline{c},\ etc.$$

Thus $\alpha^* = \alpha$ iff

$$c_{ij} = 0 \ \ if \ \ i + j \notin 2\mathbf{Z};$$
$$c_{ij} \in \mathbf{R} \ \ if \ \ i + j \in 2\mathbf{Z}.$$

Let $\alpha^* = \alpha$ and $w\alpha = \max \operatorname{supp} \alpha = x^n y^m$. Then

$$0 < \alpha \Longleftrightarrow 0 < (-1)^{(n+m)/2} c_{ij} \in \mathbf{R}.$$

defines a strong ordering which is also Baer and Jordan.

4. CONSTRUCTION OF VALUATIONS

It is possible to construct the valuation from $K \subset D$ under weaker hypotheses which do not require D to be strongly ordered. For simplicity in order to illustrate the main ideas without technicalities we simply assume D to be strongly ordered.

4.1. DEFINITION. For any division ring D, a *valuation* is a multiplicative group homomorphism $v: D\backslash\{0\} \longrightarrow \Gamma$ onto a totally ordered group Γ such that for all $a, b \in D\backslash\{0\}$

(0) $v[D\backslash\{0\}] = \Gamma$;

(i) $v(ab) = (va)(vb)$; and

(ii) $v(a + b) \leq \max[va, vb]$.

As usual, it is convenient to extend v to a semigroup homomorphism also denoted by $v: D \longrightarrow \Gamma \cup \{\overline{0}\}$ where $\overline{0}x = x\overline{0} = \overline{0} < x$ for any $x \in \Gamma$, and $v0 = \overline{0}$. Let $1 = e \in \Gamma$, and $s, t \in \Gamma$. Set $D_s = \{d \in D \mid vd < s\} \subset D^s = \{d \mid vd \leq s\}$. Assume now also that $(D, *)$ is a $*$-sfield such that $D^{e*} \subset D^e$ and $D_e^* \subset D_e$. Define $s^* = v(a^*)$ where $a \in D$ is arbitrary such that $va = s$. Then $s^{**} = s$, $(st)^* = t^* s^*$, and Γ becomes a group with an involution.

If D is a $*$-sfield, then a valuation v is a *$*$-valuation* if also $v(a^*) = va$, in which case Γ is necessarily abelian. If $(D, *, P)$ is a po-$*$-sfield, then a valuation is *compatible* with the partial order if for any $0 < a \leq b \in D\backslash\{0\}$, also $va \leq vb \in \Gamma$. Consequently, v is compatible with P if and only if $v: P \longrightarrow \Gamma$ is an order preserving function. In particular, if P is a strong order, then compatibility is equivalent to $v: (D\backslash\{0\})^{+} \longrightarrow \Gamma$ being an o-homomorphism of po-groups.

From now on it will be assumed that $K \subset D$ satisfy the following.

4.2. HYPOTHESES. Let $(D, *, P)$ be strongly ordered containing a skew subfield $K \subset D$ with $K = K^*$ satisfying the following hypotheses:

(H1) $\forall\, 0 < c \in K$, $\forall\, 0 \neq d \in D$, $\exists\, k \in K$ such that $d^{-1}cd < k$.

(H2) $\forall\, a, b \in D$, if $a^*a \leq b^*b$, then for every $0 < x \in D$, also $a^*xa \leq b^*xb$.

4.3. REMARKS. (1) $K =$ center D satisfies (H1).

(2) All examples of strongly ordered division rings $K \subset D$ in this article (e.g. $K((x;\delta))$, $K((x;\delta))((x_2;\delta_2))$, $K[[\Gamma;\theta]]$ etc.) satisfy both (H1) and (H2).

(3) All known examples of strongly ordered division rings satisfy (H2).

(4) The hypothesis (H1) should be thought of as singling out those skew subfields $K \subset D$ which are suitable for building valuations.

(5) (H1) is equivalent to:

$$\forall\, 0 < c \in K,\ \forall\, 0 \neq d \in D,\ \exists\, k \in K \text{ such that } 0 < k < d^{-1}cd.$$

PROOF. By 1.3 (g), $0 < a < b \in D$ is equivalent to $0 < b^{-1} < a^{-1}$. Apply (H1) to $0 < c^{-1} \in K$, $d^{-1} \in K$ to get $0 < dc^{-1}d^{-1} < k_1$ for some $0 < k_1 \in K$. If $k = k_1^{-1}$, then $0 < k < d^{-1}cd$. The converse is proved similarly.

4.4. QUASI-ORDER. For $a, b \in D$, define $a \prec b$ to mean that there exists a $0 < k \in K$ such that $a^*a \leq kb^*b$. Suppose that $a \prec b$ and $b \prec c$. Then $b^*b \leq k_1c^*c$, and $a^*a \leq kb^*b \leq kk_1c^*c$. Thus $a \prec c$, and "$\prec$" is a quasi-order. Consequently, it defines an equivalence relation "$\sim$" on D by $c \sim d$ if $c \prec d$ and $d \prec c$. Let $[a] = \{c \mid c \sim a\}$ denote the equivalence class of a. Define $\Gamma = \{[a] \mid 0 \neq a \in D\} \subset \Gamma \cup \{\overline{0}\}$, where $\overline{0} = [0] = \{0\}$. Let $v: D \longrightarrow \Gamma \cup \{\overline{0}\}$ be the projection $va = [a]$ onto the equivalence classes. Define $[a] \leq [b]$ if one of the following equivalent conditions holds:

(i) $a \prec b$;

(ii) $\forall\, c \in [a]$, $\forall\, d \in [b]$, $c \prec d$.

Although P is not necessarily a linear order, $(\Gamma, \leq)$ turns out to be a totally ordered set. For any $[a] \neq [b] \in \Gamma$, $b^*b - a^*a \in \text{sym}(D)\backslash\{0\} \subset P \cup -P$ by 1.1(f). Since $P \cap -P = \emptyset$, either $0 < b^*b - a^*a$ in which case $[a] \leq [b]$ or $0 < b^*b - a^*a$, and $[b] \leq [a]$.

It is not enough to merely show the existence of the valuation. If the relations "$\prec$" and "$\sim$" on D, and "$\leq$" on Γ are to be applicable and useful, some rules are needed which tell us when they hold.

4.5. PROPOSITION. For $a, b \in D$ with $b \neq 0$, the following hold.

(1) The following are all equivalent:
(i) $\exists\, 0 < k_1 \in K,\ a^*a \le k_1 b^*b$;
(ii) $\exists\, 0 < k_2 \in K,\ k_2 a^*a \le b^*b$;
(iii) $\exists\, 0 < k_3 \in K,\ a^*a \le b^*bk_3$;
(iv) $\exists\, 0 < k_4 \in K,\ a^*ak_4 \le b^*b$.

(2) The following are all equivalent:
(i) $a \prec b$;
(ii) $\exists\, 0 < k_1, k_2, k_3, k_4 \in K$ such that $k_1 a^*ak_2 \le k_3 b^*bk_4$;
(iii) $\exists\, 0 < 1 < k = k^* \in K$ such that $a^*a < kb^*b < kb^*bk$.

(3) The following are all equivalent:
(i) $[a] < [b]$;
(ii) $\forall\, 0 < k_1, k_2, k_3, k_4 \in K$;
$k_1 a^*ak_2 < k_3 b^*bk_4$. Moreover, (ii) also holds if any one, two, or three of the k_i equal 1.
(iii) $\forall\, k \in \mathrm{sym}(D) \cap K^+,\ a^*a < kb^*bk$.

PROOF. (1) (i) $\Longrightarrow$ (ii). By 1.3 (g), first $0 < k_1^{-1} \in K$ and second $k_1^{-1}a^*a \le b^*b$. Similarly (ii) $\Longrightarrow$ (i).

(1) (i) $\Longrightarrow$ (iii). Write $k_1b^*b = b^*b(b^*b)^{-1}k_1b^*b$. By (H1), $(b^*b)^{-1}k_1b^*b < k_3$ for some $k_3 \in K$. Thus $a^*a \le b^*bk_3$. The rest of the proof is omitted.

4.6. LEMMA. For any $a, b, \lambda, \rho \in D$ with $a \prec b$ the following hold
(i) $a\rho \prec b\rho$; (ii) $\lambda a \prec \lambda b$; (iii) $a^* \prec b^*$.

PROOF. By hypothesis, $a^*a \le kb^*b$ for some $0 < k \in K$. (i) Use of 1.1 (e) and 1.3 (g) gives $\rho^*a^*a\rho \le \rho^*kb^*b\rho = \rho^*k\rho^{*-1}(\rho^*b^*b\rho)$. By (H1), $0 < \rho^*k\rho^{*-1} < k_1$ for some $k_1 \in K$. Thus $(a\rho)^*a\rho \le k_1(b\rho)^*b\rho$, and $a\rho \prec b\rho$.

(ii) If $k \le 1$, then $a^*a \le b^*b$, and $(\lambda a)^*\lambda a \le (\lambda b)^*\lambda b$ by (H2). So let $1 < k$. In view of $k < k + k^*$, without loss of generality let $k = k^*$. Then multiplying $0 < 1 < k$ by kb^*b gives $0 < kb^*b \le kb^*bk$. Use of (H2) with $x = \lambda^*\lambda$ on $a^*a \le (bk)^*bk$ gives

$$a^*\lambda^*\lambda a \le kb^*\lambda^*\lambda bk = k\beta b^*\lambda^*\lambda b,$$
$$0 < \beta = (b^*\lambda^*\lambda b)k(b^*\lambda^*\lambda b)^{-1}.$$

By (H1), $\beta < k_1$ for some $k_1 \in K$. Finally, $k\beta b^*\lambda^*\lambda b < kk_1b^*\lambda^*\lambda b$. Thus $\lambda a \prec \lambda b$.

(iii) Without loss of generality, let $b \ne 0$. By 4.5 (2) (iii) we have $a^*a < (bk)^*bk$ for some $1 < k = k^*$. Application of (H2) with $x = a^{*-1}a^{-1}$ gives $0 < (bk)^*a^{*-1}a^{-1}bk$. Thus $0 < (bk^*)^{-1}(bk)^{-1} < a^{*-1}a^{-1}$. In a strongly ordered division ring, if $0 < c < d$, then $0 < d^{-1} < c^{-1}$. Thus $aa^* < bk(bk)^* = bkk^*b^{-1}bb^* < k_1bb^*$, where $0 < bkk^*b^{-1} < k_1$ for $k_1 \in K$ given by (H1). Thus $a^* \prec b^*$.

4.7. COROLLARY. $(\Gamma, \leq)$ is a totally ordered group with

(i) $[a][b] = [ab]$, $1 = e = [k]$, $[a]^{-1} = [a^{-1}]$; $a, b \in D\backslash\{0\}$; $0 \neq k \in K$.

(ii) $[a]^* = [a^*]$ defines an order preserving multiplication reversing involution on Γ.

(iii) $D_e^* = D_e$; $D^{e*} = D^e$ (See 4.1).

PROOF. (i) If $a \sim a_1$ and $b \sim b_1$, then by 4.6 (i), $ab \sim a_1b$; and by 4.6 (ii), $a_1b \sim a_1b_1$. Thus $ab \sim a_1b_1$, and Γ is a semigroup.

From $1 \leq (k^{-1}k^{-1*})k^*k$ it follows that $1 \prec k$, and from $k^*k \leq (k^*k)\cdot(1^*1)$ that $k \prec 1$. Hence $1 = e = [k] \in \Gamma$. Thus $[a]^{-1} = [a^{-1}]$.

Let $e \leq [a]$ and $e \leq [b]$. Then $k_1 \leq a^*a$ and $k_2 \leq b^*b$ for some $0 < k_1$, $k_2 \in K$ by 4.5 (1) (ii). Use of 1.1 (e) and 1.3 (g) shows that $0 < (b^*k_1b^{*-1})b^*b = b^*k_1b \leq b^*a^*ab$. Now apply 4.3 (5) to get that $0 < k_3 < b^*k_1b^{*-1}$ for some $0 < k_3 \in K$. Thus $k_3 \cdot 1 < (ab)^*ab$; by 4.5 (1) (ii), $1 \prec ab$. Thus $e \leq [a][b]$, and Γ is a totally ordered group.

(ii) By 4.6 (iii), $[a^*] = \{c^* \mid c \sim a\} = [a]^*$ is a well defind order preserving anti-automorphism of Γ.

(iii) Since $e^* = e$, again use of 4.6 (iii) shows that $D_e^* \subseteq D_e$, and hence $D_e^{**} \subseteq D_e^*$. Thus $D_e^* = D_e$ and $D^{e*} = D^e$.

4.8. THEOREM. Let $(D, *, P)$ be a strongly ordered division ring (1.1 (a)-(f)) and $K = K^* \subset D$ any skew subfield satisfying conditions 4.2 (H1) and (H2). The set

(i) $\Gamma = \{[a], [b], \ldots \mid a, b \in D\backslash\{0\}\}$ is a totally ordered group.

(ii) $v: D \longrightarrow \Gamma \cup \{\bar{0}\}$, $va = [a]$, $v0 = \bar{0}$, $v(D\backslash\{0\}) = \Gamma$, is a valuation.

(iii) v is compatible with the partial order on D; the restriction $v: D^+\backslash\{0\} \longmapsto \Gamma$ is an order preserving homomorphism of the po-group $D^+\backslash\{0\}$ onto the o-group Γ.

(iv) Γ is a group with an order preserving involution $[a]^* = [a^*]$.

(v) For any $0 \neq k \in K$, $v(ka) = v(ak) = v(a)$, $v(k) = e$; $(a + D_e)^* = a^* + D_e$ defines an involution on the residue field D^e/D_e of v; $K \cong (K + D_e)/D_e \subset D^e/D_e$ under an involution preserving canonical isomorphism.

The author thanks the referees for a careful reading of both papers and for suggesting some improvements.

REFERENCES

[Ch 1] M. Chacron, C-orderable division rings with involution, J. Algebra 75 (1982), 495-521.

[Ch 2] M. Chacron, C-valuations and normal c-orderings, Canad. J. Math., to appear.

[Ch 3] M. Chacron, Non-isotropic unitary spaces and modules with Cauchy-Schwarz inequalities, Pac. J. Math. 119 (1985), 1-87.

[Co 1] P. Conrad, On ordered division rings, Proc. Amer. Math. Soc. (1954), 323-328.

[Co 2] P. Conrad, Generalized semigroup rings, Jour. Indian Math. Soc. 21 (1957), 73-95.

[CD] P. Conrad and J. Dauns, An embedding theorem for lattice ordered fields, Pac. J. Math. 30 (1969), 385-398.

[Cr 1] T. Craven, Orderings and valuations on ∗-fields, to appear.

[Cr 2] T. Craven, Approximation properties for orderings on ∗-fields, to appear.

[Cr 3] T. Craven, Witt groups of Hermitian forms over ∗-fields, to appear.

[D1] J. Dauns, A Concrete Approach to Division Rings, Research and Education in Math. 2 (1982), Heldermann Verlag, Berlin.

[D2] J. Dauns, Mathiak valuations, in: Ordered Algebraic Structures, Marcel Dekker, 1985; pp. 33-75.

[D3] J. Dauns, Metrics are Clifford algebra involutions, International J. Theoretical Physics 27 (1988), 183-191.

[D4] J. Dauns, Centers of semigroup rings, Semigroup Forum, 38 (1989), 355-364.

[D5] J. Dauns, Lattice ordered division rings, in preparation.

[GP] H. N. Gupta and A. Prestel, On a class of Pasch-free Euclidean planes, Bull. Acad. Polon. Sci. Ser. Sci. Math. Astronom. Phys. XX (1972), 17-23.

[H1] S. Holland, Jr., Ordering and square roots in ∗-fields, J. Algebra 46 (1977), 207-219.

[H2] S. Holland, Jr., ∗-valuations and ordered ∗-fields, Trans. Amer. Math. Soc. 262 (1980), 219-243.

[H3] S. Holland, Jr., Strong ordering of ∗-fields, J. Algebra 101 (1986), 16-46.

[H4] S. Holland, Jr., Erratum to "∗-valuations and ordered ∗-fields", Trans. Amer. Math. Soc. 262 (1980), 219-243.

[I] I. Idris, $*$-valuated division rings, orderings and elliptic hermitian spaces, Ph.D. thesis, Carleton University, Ottawa, Canada (1986).

[Mc] R. McHaffey, A proof that the quaternions do not form a lattice ordered algebra, Proc. Iraqi Sci. Soc. 5 (1962), 670-671.

[P1] A. Prestel, Quadratische Semi-Ordnungen und quadratische Formen, Math. Z. 133 (1973), 319-342.

[P2] A. Prestel, Lectures on formally real fields, Springer Lecture Notes No. 1093, Springer-Verlag, Berlin, 1984.

[R] R. H. Redfield, Embeddings into power series rings,

[S] S. Steinberg, An embedding theorem for commutative lattice-ordered domains, Proc. Amer. Math. Soc. 31 (1972), 409-416.

LATTICE ORDERED DIVISION RINGS EXIST

John Dauns
Department of Mathematics
Tulane University
New Orleans, LA 70118

A noncommutative lattice ordered division ring D which is not totally ordered here will be called a *proper ℓ-division ring.* So far in the literature, there are two general methods of constructing ℓ-division rings, and one example. An easily applicable and clear method is in [CD; p.387, Proposition 2.2]; and a generalization of a special case of this is in [D2; p.304, VII-2.9, VII-2.11]. The latter is used to construct an actual example in [D2; p.318, VII-4.4]. However, in the latter algebraic and lattice operations cannot be carried out explicitly because the Ore condition is used in the construction.

The class of proper ℓ-division rings D constructed here are group power series division rings. ([D3; p.54] or [D2; p.308]) of the form $K[[\Gamma]]$, or $K[[\Gamma;\theta]]$, where K is a totally ordered skew field ([D2; p.317] or [D1; p.242]), and Γ is a partially ordered group. In all examples in this note $1 > 0$, and the lattice order on D extends to a total order. Whether these hold for all ℓ-division rings are two open questions. The techniques for constructing finite dimensional commutative ℓ-fields ([W] and [CD]) do not seem to generalize. It is a third open question whether there exists a proper ℓ-division ring which is finite dimensional over its center (see [Mc]). The answer is yes iff the answer to the second question is no, because by Albert's theorem ([D2; p.294]), the lattice order of a finite dimensional proper ℓ-division ring (if it exists) can not be extended to a total order.

The existence of a proper ℓ-division ring D, in which the lattice and algebraic operations can be concretely and transparently performed, was posed as an open problem at the Workshop-Conference on Ordered Algebraic Structures at Curaçao in August, 1988.

J. Martinez (ed.), Ordered Algebraic Structures, 229–234.

1. Construction

A class of proper ℓ-division rings is constructed by generalizing and unifying the procedures [CD; p.387, Proposition 2.2] and [D2; p.304-305, VII-2.9 and VII-2.11; p.318, VII-4.4].

1.1. Twisted Group Power Series Division Rings

Let Γ be a (torsion free in general noncommutative) group which has a total group order $\leq_1$, so that $(\Gamma, \leq_1)$ is an o-group. Let K be a totally ordered skew field, and $\theta: \Gamma \longrightarrow o - \text{End } K$ a semigroup homomorphism into the semigroup of all order preserving (monic) endomorphisms of $K \longrightarrow K$. Then automatically $\Gamma\theta \subseteq o - \text{Aut } K$, and $\theta: \Gamma \longrightarrow o - \text{Aut } K$ is a group homomorphism into the (unordered) group of all o-automorphisms of K.

For any $x, y \in \Gamma$ and any $k, c \in K$:

$$x\theta: K \longrightarrow K,\ (x\theta)K^+ = K^+,\ (x^{-1}\theta)K^+ = K^+,\ e \equiv 1 \in \Gamma,$$
$$e\theta = 1_K,\ k^x \equiv k^{x\theta} = k(x\theta),\ (k^x)^y = k^{xy},$$
$$(k - c)^x = k^x - c^x,\ (kc)^x = k^x c^x,\ k^{-x} \equiv 1/k^x = (k^{-1})^x$$

Define the *twisted group power series division* ring $K[[\Gamma; \theta]]$ as the set of all formal sums $\alpha = \sum\{x\alpha(x) \mid x \in \Gamma\}$, $\beta = \sum\{y\beta(y) \mid y \in \Gamma\}$; $\alpha(x), \beta(y) \in K$, whose *supports* $\text{supp } \alpha = \{x \in \Gamma \mid 0 \neq \alpha(x) \in K\}$ satisfy the ascending chain condition in $\text{supp } \alpha \subset (\Gamma, \leq_1)$. Addition is pointwise, $kx = xk^x$ for $k \in K$, and $\alpha\beta = \sum\{xy\alpha(x)^y\beta(y) \mid (x, y) \in \Gamma \times \Gamma\}$. (See [D3; p.54] or [D2; p.308]).

Then $K[[\Gamma; \theta]]$ is a totally ordered division ring under the lexicographic order with $0 \leq \alpha$ if $0 \leq \alpha(x)$ where $x = \max \text{ supp } \alpha$ is the largest element of the support of α. Later, $K[[\Gamma; \theta]]$ will be also lattice ordered.

If $x\theta = 1_K$ for all $x \in \Gamma$, write $K[[\Gamma]] \equiv K[[\Gamma; 1_K]]$. If $(\Gamma, \leq_1) = \{\cdots x^{-2} < x^{-1} < e = 1 < x < \cdots\}$, and $\psi: \Gamma \longrightarrow o\text{-}\text{Aut } K$, abbreviate $K[[\Gamma; \psi]] = K((x; \theta))$ where $\theta = x\psi$. Call $K((x; \theta))$ the *skew Laurent power series division ring*. If $\theta = 1_K$, write $K((x)) = K((x; 1_K))$.

1.2. Groups

Suppose that Γ is a torsion free group and $H \triangleleft \Gamma$ a normal totally ordered subgroup such that $H^+ \triangleleft \Gamma$. Also assume that there exists a total group order $\leq_1$ on Γ extending that of H (See [NS], [MR], and [RH] for large classes of such Γ.) Let $\Gamma_1 = (\Gamma, \leq_1)$ be this o-group with the same underlying set of elements as Γ.

Let H have its given total order, but in addition for any $g_1, g_2 \in \Gamma \backslash H$ and $h_1, h_2 \in H$, define $g_1h_1 \leq g_2h_2$ iff $g_1 = g_2$ and $h_1 \leq h_2 \in H$. This defines a po-group structure $\langle\Gamma, \leq\rangle$ on Γ. Since $H^+ \triangleleft \Gamma$, the total order on Γ_1 extends this po-group order $\langle\Gamma, \leq\rangle$. An important special case is when $H = \{e\} = 1$, in which case $\langle\Gamma, \leq\rangle$ is discrete. Note that $\langle\Gamma, \leq\rangle$ is an ℓ-group iff $\Gamma = H$.

1.3. Lattice Order

For K, θ, as in 1.1 and Γ as above in 1.2, define $K[[\Gamma;\theta]]$ as $K[[\Gamma;\theta]] = K[[\Gamma_1;\theta]]$, where in the latter $\Gamma = \Gamma_1$ as a set, but with the extended total order $\Gamma_1 = (\Gamma, \leq_1)$. For $\alpha \in K[[\Gamma,\theta]]$, max supp α denotes the set of maximal elements of supp α in the partial order $\langle \Gamma, \leq \rangle$. Define $0 \leq \alpha$ iff $0 \leq \alpha(x) \in K$ for every $x \in$ max supp α. For each $x \in \max\operatorname{supp} \alpha$, define $\alpha_{[x]} \in K[[\Gamma;\theta]]$ by $\alpha_{[x]} = \sum\{y\alpha(y) \mid y \leq x\}$, and $\alpha^+ = \sum\{\alpha_{[x]} \mid x \in \max\operatorname{supp} \alpha,\ 0 < \alpha(x)\}$, $\alpha^- = -\sum\{\alpha_{[x]} \mid x \in \max\operatorname{supp} \alpha,\ \alpha(x) < 0\}$. Thus $\alpha = \alpha^+ - \alpha^-$, $\alpha^+ \wedge \alpha^- = \operatorname{glb}(\alpha^+, \alpha^-) = 0$, and lub $\{\alpha, 0\} = \alpha \vee 0 = \alpha^+$ exist. Consequently $K[[\Gamma;\theta]]$ is an additive ℓ-group (See [C2; pages 0.3 and 0.12]). Since max supp $\alpha\beta \subseteq$ (max supp α)·(max supp β) and since $K^+y = yK^+$ for all $y \in \Gamma$, it follows that when $0 < \alpha = \alpha^+$ and $0 < \beta = \beta^+$, then also $\alpha\beta = \alpha^+\beta^+ = (\alpha\beta)^+ > 0$. Thus $K[[\Gamma,\theta]]$ is a lattice ordered division ring. It is not totally ordered iff $H \neq \Gamma$, and noncommutative if any one of the following hold: (i) $x\theta \neq 1$ for some $x \in \Gamma$; (ii) K is noncommutative; or (iii) Γ is nonabelian.

2. Applications

The above general method now is used to construct actual concrete examples where the lattice and algebraic operations can be given by explicit formulas.

2.1. Example

Take any noncommutative totally ordered skew field K (e.g. $K = \mathbf{R}(t)((x;\theta))$ as in 2.2 below - but reordered with the total lexicographic order). In 1.3, take $\Gamma = \{\ldots, x^{-1}, 1, x, \ldots\}$ unordered and $H = \Gamma^{(3)} = \{\cdots < x^{-6} < x^{-3} < 1 < x^3 < x^6 < \cdots\}$. Then 1.3 (ii) makes $K((x))$ into a proper ℓ-division ring. Each $\alpha \in K((x))$ is of the form $\alpha = \alpha_0 + \alpha_1 + \alpha_2$, $\alpha_0 = \sum\{x^{3i}k_{3i} \mid i \leq m(0)\}$, $\alpha_1 = \sum\{x^{3i+1}k_{3i+1} \mid i \leq m(1)\}$, $\alpha_2 = \sum\{x^{3i+2}k_{3i+2} \mid i \leq m(2)\}; m(0), m(1), m(2) \in Z; k_i \in K$. Thus $0 \leq \alpha$ iff $0 \leq k_{3m(0)}$, $0 \leq k_{3m(1)+1}$, $0 \leq k_{3m(2)+2}$. Clearly, $\Gamma \cong \mathbf{Z}$ above can be replaced by any free abelian group, and "3" by any positive integer ≥ 2.

The next three interrelated examples show the usefulness of groups of the form $o-$ Aut K, where K is a totally ordered division ring.

2.2. Example

For $\mathbf{R}$ the reals, $\mathbf{R}(t) \subset \mathbf{R}((t))$ are the fields of rational functions and the commutative field of all formal Laurent series with only a finite number of positive exponent terms, totally ordered lexicographically as in 1.1. It is known ([J; p.514-515]) that any arbitrary field automorphism θ of $\mathbf{R}(t)$ keeping $\mathbf{R}$ elementwise fixed is of the form $f(t)\theta = f(t\theta)$, where $f(t) \in \mathbf{R}(t)$ and $t\theta = (at+b)/(ct+d)$ with $a, b, c, d \in \mathbf{R}$, and $ad - bc \neq 0$. Every o-automorphism of $\mathbf{R}(t)$ is the identity on $\mathbf{R}$. If $c \neq 0$, then for $0 < M \in \mathbf{R}$ sufficiently large, $0 < -M + t$ but $-M + t\theta < 0$. Thus every element θ of $o -$ Aut $\mathbf{R}(t)$ is precisely of the form $t\theta = at + b$ with b, $0 < a \in \mathbf{R}$. In 1.2, take $H = \{x^0\} = 1$ and $\Gamma = \{\ldots, x^{-1}, 1, x, \ldots\}$ discretely ordered.

Form the skew Laurent power series division ring $\mathrm{R}(t)((x;\theta))$, where $fx = xf^{\theta}$. By 1.4 (i), this is a proper ℓ-division ring. For $\alpha = \sum\{x^i f_i \mid i \leq n\} \in \mathbf{R}(t)((x;\theta))$, $n \in \mathbf{Z}$, $f_i \in \mathbf{R}(t)$, $\alpha \vee 0 = \sum x^i(f_i \vee 0)$, and $0 \leq \alpha$ iff all $f_i \geq 0$. Similarly $\mathbf{R}((t))((x;\theta))$ is also a proper ℓ-division ring.

2.3. Example

For the reals $\mathbf{R}$ and $\mathbf{R}^+ = \{a \mid 0 < a \in \mathbf{R}\}$ with the natural order, let $\Gamma = \mathbf{R}^+ \times \mathbf{R}$ be the semidirect product totally ordered lexicographically:

$$(y,z)(\overline{y},\overline{z}) = (y\overline{y},\ y\overline{z} + z) \qquad y, \overline{y} \in \mathbf{R}^+,\ z, \overline{z} \in \mathbf{R};$$

$$(y,z) \leq_1 (\overline{y},\overline{z}) \text{ if either } y \leq \overline{y}; \text{ or if } y = \overline{y} \text{ but } z \leq \overline{z};$$

$$(y,z)^{-1}(\overline{y},\overline{z})(y,z) = (\overline{y}, y^{-1}\overline{y}z + y^{-1}\overline{z} - y^{-1}z).$$

The above identities show that $\Gamma_1^+ \triangleleft \Gamma$. In 1.3, take $H = \{1\} \times \mathbf{R} \triangleleft \Gamma$. Take any totally ordered commutative field such as the rationals $\mathbf{Q}$. Then $D = \mathbf{Q}[[\Gamma]]$ by 1.4 (iii) is a proper ℓ-division ring. For $\alpha = \sum\{(y,z)\alpha(y,z) \mid (y,z) \in \mathbf{R}^+ \times \mathbf{R}\} \in D$, $\alpha(y,z) \in \mathbf{Q}$, set $\alpha_y = \sum\{(y,z)\alpha(y,z) \mid z \in \mathbf{R}\}$. Thus $\alpha = \sum\{\alpha_y \mid y \in \mathbf{R}^+\}$, $\alpha \vee 0 = \sum\{\alpha_y \mid y \in \mathbf{R}^+,\ 0 < \alpha\ (\max \operatorname{supp} \alpha_y)\}$, and $0 \leq \alpha$ iff $0 \leq \alpha(\max \operatorname{supp} \alpha_y)$ for all y.

2.4. Example

For $\varphi \in o -$ Aut $\mathbf{R}(t)$ as in 2.2, the definition $1 \leq \varphi$ iff $f \leq f\varphi$ for all $f \in \mathbf{R}(t)$ makes $o -$ Aut $\mathbf{R}(t)$ into a naturally ordered o-group. For Γ as in 2.3, $\theta: \Gamma \longrightarrow o-$Aut $\mathbf{R}(t)$ is an o-isomorphism of o-groups, where $(y,z)\theta: \mathbf{R}(t) \longrightarrow \mathbf{R}(t)$ by $f(t)[(y,z)\theta] = f(ty+z)$ for $(y,z) \in \Gamma$ and $f(t) \in \mathbf{R}(t)$. Take $H = \{1\} \times \mathbf{R} \triangleleft \Gamma$ as before in 2.3. Then $D = \mathbf{R}(t)[[\Gamma;\theta]]$ again is a proper ℓ-division ring by not only 1.4 (iii) but also by 1.4 (i). For $\alpha = \sum\{(y,z)\alpha(y,z) \mid (y,z) \in \mathbf{R}^+ \times \mathbf{R}\} \in D$, the rest is as before except that now $\alpha(y,z) \in \mathbf{R}(t)$. Note that $\mathbf{Q}[[\Gamma]] = \mathbf{Q}[[\Gamma,\theta]] \subset \mathbf{R}(t)[[\Gamma;\theta]]$, where the former is as in 2.3.

ACKNOWLEDGEMENT

During the Curaçao conference Robert Redfield told me that he thought he could construct proper ℓ-division rings based on B. Neumann's paper [N]. Although I did not understand his comments (partly because [N] is a difficult paper), nevertheless later at the same conference P. Conrad told me that R. Redfield had rediscovered the proper ℓ-division ring [D2; p.318, VII-4.4].

Robert Redfield's insight and enthusiasm has been a catalyst. The author also thanks Paul Conrad for his interest and suggestions in constructing proper ℓ-division rings.

REFERENCES

[C1] P. Conrad, On ordered division rings, Proc. Amer. Math. Soc. (1954), 323-328.

[C2] P. Conrad, Lattice Ordered Groups, Tulane Lecture Notes, 1970.

[CD] P. Conrad and J. Dauns, An embedding theorem for lattice ordered fields, Pac. J. Math. 30 (1969), 385-398.

[CHH] P. Conrad, J. Harvey, and C. Holland, The Hahn embedding theorem for abelian lattice ordered groups, Trans. Amer. Math. Soc. 108 (1963), 143-169.

[D1] J. Dauns, Generalized semigroup rings, in: Algebra Carbondale 1980, Springer Lecture Notes in Math. 848 (1981), pp. 235-254.

[D2] J. Dauns, A Concrete Approach to Division Rings, Heldermann Verlag, Berlin 1982.

[D3] J. Dauns, Mathiak valuations, in: Ordered Algebraic Structures, Marcel Dekker, 1985; pp. 33-75.

[D4] J. Dauns, Lattice ordered division rings, in preparation.

[F] L. Fuchs, Partially Ordered Algebraic Systems, Pergammon Press, Addison-Wesley, Reading, Mass., 1963.

[J] N. Jacobson, Basic Algebra II, W. H. Freeman and Co., San Francisco, 1980.

[N] B. Neumann, On order division rings, Trans. Amer. Math. Soc. 66 (1949), 202-252.

[NS] B. Neumann and J. Sheppard, Finite extensions of fully ordered groups, Proc. Royal Soc. London 239 (1957),320-327.

[Mc] R. McHaffey, A proof that the quaternions do not form a lattice ordered algebra, Proc. Iraqi. Sci. Soc. 5 (1962), 670-671.

[MR] R. Mura and A. Rhemtulla, Orderable Groups, Marcel Dekker, New York, 1977.

[RE] R. H. Redfield, Embeddings into power series rings, to appear.

[RH] A. Rhemtulla, Periodic extensions of ordered groups, to appear, these proceedings.

[S] S. Steinberg, An embedding theorem for commutative lattice-ordered domains, Proc. Amer. Math. Soc. 31 (1972), 409-416.

[W] R. Wilson, Lattice orderings on the real field, Pac. J. Math. 63 (1976), 571-577.

VECTOR-LATTICES AND A PROBLEM IN GEOMETRY

by James J. Madden, Indiana University at South Bend

Introduction: Let X be a subset of $\mathbb{R}^n$ and suppose $f:X \to \mathbb{R}$ is "weakly piecewise linear" in the following sense: there is a finite collection of closed piecewise linear sets $\{P_i\}$ such that $X \subseteq \cup P_i$ and for each i, the restriction of f to $X \cap P_i$ is (affine) linear. Is f "strictly piecewise linear", i.e., is f the restriction to X of a piecewise linear function defined on $\mathbb{R}^n$? In general the answer is no, though it is not hard to think of conditions on X which are sufficient for a yes answer, e.g., X itself is a closed p.l. set. We show in this paper that there is a geometric condition on the embedding $X \subseteq \mathbb{R}^n$ which is necessary and sufficient for every weakly p.l. $f:X \to \mathbb{R}$ to be strictly p.l.. It is a kind of local connectedness near the points of the remainder of a certain non-Hausdorff completion of X. The proof involves facts about vector-lattice ideals and, implicitly, the duality between finitely presented vector-lattices and piecewise linear sets and maps due to Beynon (1975) and anticipated by Baker (1968). While the Baker-Beynon theory has been used in the past to derive results about vector-lattices from well known geometric facts, the present paper is the first instance known to the author in which it is used to derive a theorem in elementary geometry from algebraic facts about vector-lattices.

J. Martinez (ed.), Ordered Algebraic Structures, 235–245.

1. Definitions. A closed half-space in $\mathbb{R}^n$ is a set of the form $\{x \mid f(x) \geq 0\}$, where $f:\mathbb{R}^n \to \mathbb{R}$ is linear. Here and below, "linear" is taken in the affine sense. A closed convex p.l. set is an intersection of finitely many closed half-spaces and a closed p.l. set is a union of finitely many closed convex p.l. sets. A function $f:\mathbb{R}^n \to \mathbb{R}$ is p.l. if its graph is a closed p.l. set. The weakly and strictly p.l. functions on $X \subseteq \mathbb{R}^n$ are defined as in the introduction. If $X \subseteq \mathbb{R}^n$, the set of all weakly p.l. functions on X is a vector-lattice under the pointwise operations which is denoted $WPL(X)$. The strictly p.l. functions on X also form a vector-lattice $PL(X) \subseteq WPL(X)$.

Examples. 1) If $X = \mathbb{R} - \{0\} \subseteq \mathbb{R}$, then $WPL(X)$ includes the function $x/|x|$ which is not strictly p.l. If $Y = \mathbb{R} - [-1,1] \subseteq \mathbb{R}$, then $WPL(Y) = PL(Y)$. This draws attention again to the fact that our problem has to do with the nature of the embedding of X in $\mathbb{R}^n$ rather than just with intrinsic properties of the set X.

2) Let $X = \{(x,y) \in \mathbb{R}^2 \mid |y| = x^2\}$ and let $h:X \to \mathbb{R}$ be defined by $h(x,y) = x$ if $y \geq 0$ and by $h(x,y) = 0$ if $y \leq 0$. Then h is weakly p.l.. It is not strictly p.l. because at $(0,0)$ it is not Lipschitz continuous. Indeed, $\lim_{x\to 0^+} (h(x,x^2) - h(x,-x^2))/\|(x,x^2) - (x,-x^2)\| = +\infty$. On the other hand, let $Y = \{(x,y) \in \mathbb{R}^2 \mid y = x^2 \text{ or } y = 0\}$. Then $WPL(Y) = PL(Y)$. Even though $(0,0)$ is a singularity of Y, it does not cause any problem. Essentially, the reason is that any closed p.l. set containing a neighborhood of $(0,0)$ on the curve $y = x^2$ also contains a neighborhood of $(0,0)$ on the line $y = 0$.

3) If $X \subseteq \mathbb{R}^n$ is a smooth compact curve then $WPL(X) = PL(X)$. This is because if $X \subseteq P_i$, then there are closed p.l. sets $P_i' \subseteq P_i$ such

that $X \subseteq \bigcup P_j'$ and when $i \neq j$ $P_i' \cap P_j'$ is either empty or comprises a single point of X.

4) If $X = \mathbb{R}^2 - \{(0,0)\} \subseteq \mathbb{R}^2$, then $WPL(X) = PL(X)$. Indeed, any weakly p.l. function on X has a unique continuous extension to $\mathbb{R}^2$ and this extension is p.l.

2. <u>Geometric ℓ-ideals</u>. Suppose $X \subseteq \mathbb{R}^n$. Let $C(X)$ denote the vector-lattice of all continuous real-valued functions on X. Let $V(X) \subseteq C(X)$ be a sub-vector-lattice which contains the constants and separates the points. If $f \in V(X)$, $Z(f)$ denotes $\{x \in X \mid f(x) = 0\}$. A prime ℓ-ideal $p \subseteq V(X)$ will be called <u>geometric</u> if $g \in p$ whenever $Z(g) \supseteq Z(f)$ and $f \in p$. This is essentially the same notion as the prime z-ideals of Gillman and Jerison (1960), but relativized to $V(X)$. A <u>$V(X)$-zero-set</u> is a set of the form $Z(f) \subseteq X$, where $f \in V(X)$. Obviously the geometric primes of $V(X)$ are in one-to-one correspondence with the prime filters of $V(X)$-zero-sets.

Note that the determination of whether a maximal ℓ-ideal in $V(X)$ is geometric depends on X, and not just the algebraic structure of $V(X)$. For example, every maximal ℓ-ideal of $PL([0,1])$ is geometric, but $PL((0,1])$, which is isomorphic to $PL([0,1])$, has a maximal ℓ-ideal which is not geometric. This dependency, however, occurs <u>only</u> for maximal ℓ-ideals. If $p \subseteq V(X)$ is a prime ℓ-ideal which is <u>not</u> maximal, then whether or not p is geometric will be determined by the isomorphism type of $V(X)$. We do not use this fact in this paper, but we do use the following important

<u>Lemma 2.1.</u> Any minimal prime ℓ-ideal of $V(X)$ is geometric.

Proof: From the well-known characterization of minimal primes, if p is minimal and $f \in p$, then there is $h \in V-p$ such that $|f| \wedge |h| = 0$. If $Z(g) \supseteq Z(f)$, then $|g| \wedge |h| = 0$, so $g \in p$. □

Now we specialize the discussion to WPL(X) and PL(X).

Lemma 2.2: The WPL(X)-zero-sets and the PL(X)-zero-sets coincide. They are precisely the sets of the form $K \cap X$ where K is a closed p.l. set. Hence, there is a one-to-one correspondence between the geometric prime ℓ-ideals of WPL(X) and those of PL(X).

Proof: It is immediate from the definition of a weakly p.l. function that $Z(f)$ is the intersection of X with a closed p.l. set whenever f is weakly p.l.. Since any closed p.l. subset of $\mathbb{R}^n$ is the zero set of some p.l. function, the first two assertions are clear. Since the geometric primes correspond to minimal prime filters of WPL(X)-zero-sets, the last assertion follows. □

Lemma 2.3: Let p be a prime ℓ-ideal of WPL(X), and let $f \in$ WPL(X). Then there is an affine linear function $a: \mathbb{R}^n \to \mathbb{R}$ such that $f + p = a + p$. It follows that $q \longrightarrow q \cap$ PL(X) is a bijection between the primes of WPL(X) containing p and the primes of PL(X) containing $p \cap$ PL(X).

Proof: It is enough to prove this when p is a minimal prime. Suppose $f = a_i$ on $P_i \cap X$, $i = 1, \ldots, m$, with a_i affine linear and $\cup\{P_i \mid i = 1, \ldots, m\} \supseteq X$. Since $\{Z(g) \mid g \in p\}$ is a prime filter, there is P_i such that $X \cap P_i$ is the zero set of some $g \in p$. Therefore $f - a_i \in p$, since p is geometric. This proves the first assertion, and shows that WPL(X)/p is finite dimensional and is isomophic to PL(X)/p $\cap$ PL(X). The second assertion is then apparent. □

Lemmas 2.2 and 2.3 give a clear picture of the relationship between the primes of WPL(X) and the primes of PL(X) . As commonly done, we may depict the primes of these vector-lattices as the nodes on (inverted) trees, with the nodes further from the terminus of a branch (i.e., higher up in the picture) denoting larger primes. The map $q \to q \cap PL(X)$ from the primes of WPL(X) to the primes of PL(X) is bijective on any branch and bijective on the terminii, but due to possible "earlier branching" in the primes of WPL(X) , a given non-minimal prime in PL(X) may have several pre-images. The terminology which is sometimes used is that the primes of WPL(X) are "more stranded" than the primes of PL(X) .

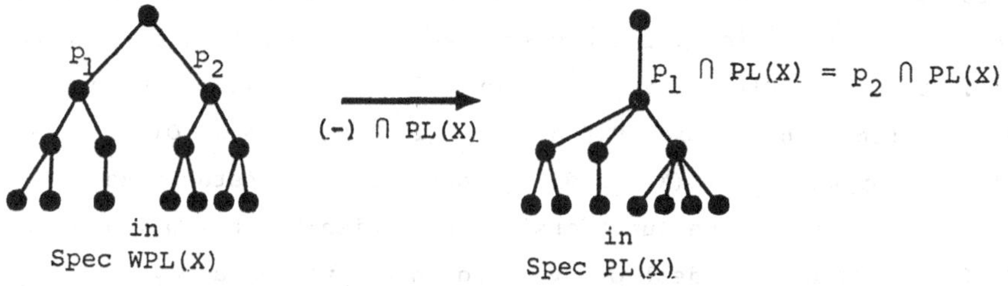

Anderson and Conrad (1981) prove a theorem more general than the following: Suppose G is an ℓ-subgroup of an abelian ℓ-group H . If the map $q \to q \cap G$ from the primes of H to the primes of G is a bijection and if $p + G = H$ for each prime $p \subseteq H$, then $G = H$. From this and lemma 2.3 it is immediate that

<u>Proposition 2.4</u>: $WPL(X) = PL(X)$ if and only if for each prime $p \subseteq PL(X)$ there is no more than one prime $q \subseteq WPL(X)$ such that $q \cap PL(X) = p$.

<u>3. Cuts and splits</u>. Michael (1964) studies the following notion of "cutting". Let S be a topological space, let $X \subseteq S$ be a dense subspace and let $p \in S - X$. One says that X <u>is cut near</u> p (more precisely, (S,X) is cut near p) if p has an open neighborhood

$U \subseteq S$ such that $U \cap X$ is a disjoint union of two (or more) non-empty open sets (of X) each with p in its closure. For example, $(\mathbb{R}, \mathbb{R} - \{0\})$ is cut near 0, but $(\mathbb{R}^2, \mathbb{R}^2 - \{(0,0)\})$ is not cut, c.f. examples 1 and 4 in §2. We relativize this notion in order to apply it to our problem. If S is equipped with a distinguished basis $\mathcal{D}$ for its open sets, then we say X <u>splits</u> near p (more precisely, $(S,\mathcal{D},X)$ splits near p) if p has an open neighborhood $U \subseteq S$ such that $U \cap X$ is a disjoint union of two or more non-empty sets of the form $D_i \cap X$, with $D_i \in \mathcal{D}$ and $p \in cl(D_i \cap X)$. (The definition is not changed if we demand that $U \in \mathcal{D}$).

Let V be a vector-lattice. Recall that Spec V is the topological space having as its points the prime ℓ-ideals of V and having as a basis for its opens the sets $D(f) = \{p | f \notin p\}$, where $f \in V^+$. The sets $D(f)$, $f \in V^+$, actually form a lattice under union and intersection because $D(f \wedge g) = D(f) \cap D(g)$ and $D(f \vee g) = D(f) \cup D(g)$. Suppose $V \subseteq C(X)$ and V contains the constants and separates points. The functions in V vanishing at a given point of X form a maximal ℓ-ideal of V, and this gives us a way of identifying X with a subset of Spec V. X is dense in Spec V and the topology which X inherits from Spec V is the same as that originally given. Later, in the proof of 3.1, we use the fact that if $f, g \in V$, and $p \subseteq V$ is a prime and $f < g$ (mod p), then $f < g$ (mod p) for all q in a sufficiently small neighborhood of p.

Let $X \subseteq \mathbb{R}^n$, and consider Spec PL(X) with the distinguished basis $\mathcal{D} = \{D(f) | f \in PL(X)^+\}$. Our main result is the following

<u>Theorem 3.1</u>: WPL(X) = PL(X) if and only if (Spec PL(X), $\mathcal{D}$, X) splits near no point of Spec PL(X) - X.

Before giving the proof below, we shall make some preliminary remarks. The first thing to be pointed out is that in proving 3.1, we may assume without loss of generality that X is bounded. To see this, identify $\mathbb{R}^n$ with $\{1\} \times \mathbb{R}^n \subseteq \mathbb{R}^{n+1}$, and project radially (i.e., along lines through the origin in $\mathbb{R}^{n+1}$) to the unit p.l. sphere $\Sigma^n \subseteq \mathbb{R}^{n+1}$. Even though the projection is not a p.l. map, it induces an isomorphism between $PL(\mathbb{R}^n)$ and $PL(\Sigma^n_+)$ where Σ^n_+ is the top half of the sphere, and between the weakly p.l. functions on X and the weakly p.l. functions on the projection of X. If X is bounded, then there is no loss of generality in assuming, as we do below, that $X \subseteq [-1,1]^n$.

In this paragraph, we introduce the symbols Aff(p) and h(p). If $Z \subseteq \mathbb{R}^n$, let Aff(Z) denote the affine subspace of $\mathbb{R}^n$ spanned by Z. (Thus, if Z is a single point, Aff(Z) = Z). Let p be any geometric prime ℓ-ideal of PL(X) or of WPL(X). We let Aff(p) denote the intersection of all Aff(Z) where Z varies over the zero sets of elements of p, i.e., $Aff(p) = \bigcap\{Aff(Z(f)) \mid f \in p\}$. Now suppose $h \in WPL(X)$. It is possible to choose $f \in p$ such that $Z(f) \subseteq Aff(p)$ and h agrees with a linear function on Z(f) (Lemma 2.3). Since p is geometric, Z(f) must contain a spanning set for Aff(p). Thus, h determines an affine linear function on Aff(p). This function we call h(p).

Aff(p) and h(p) can also be defined when p is not geometric. Restriction to X of p.l. functions on $[-1,1]^n$ provides a surjective ℓ-homomorphism $\pi : PL([-1,1]^n) \to PL(X)$. By Baker (1968), every prime of $PL([-1,1]^n)$ is geometric. If $p \subseteq PL(X)$ is any prime, we let Aff(p) denote $Aff(\pi^{-1}(p))$. If $p \subseteq WPL(X)$ is any prime, we let Aff(p) denote $Aff(p \cap PL(X))$. We leave it to the reader to check that these definitions are consistent with those of the previous paragraph when p is geometric. The reader should also note that $Aff(p) \subseteq Aff(q)$ whenever $p \supseteq q$.

If $p \subseteq PL(X)$ is any prime and $h \in PL(X)$, we define h(p) in the following manner. Choose any geometric prime $q \subseteq PL(X)$ such that $q \subseteq p$. Let $h(p) = h(q)|\mathrm{Aff}(p)$ (= $h(q)$ restricted to $\mathrm{Aff}(p)$). For any prime $p \subseteq \mathrm{WPL}(X)$ and $h \in \mathrm{WPL}(X)$, we similarly define $h(p) = h(q)|\mathrm{Aff}(p)$ where $q \subseteq \mathrm{WPL}(X)$ is any geometric prime contained in p. (When working in $\mathrm{WPL}(X)$, it is not adequate to choose q such that $q \cap \mathrm{PL}(X) \subseteq p$, since this does not guarantee $q \subseteq p$.) Again, we leave it to the reader to verify that $h(p)$ is always well-defined (i.e., does not depend on choice of q) and that the extended definition of $h(p)$ is therefore consistent with that given above for geometric p. Thus, we have the following relation for any two primes q and p of $\mathrm{PL}(X)$ (resp. $\mathrm{WPL}(X)$) and $h \in \mathrm{PL}(X)$ (resp. $h \in \mathrm{WPL}(X)$): if $q \subseteq p$ then $h(q)|\mathrm{Aff}(p) = h(p)$.

Here are some examples to illustrate the preceeding definitions. First, if p is a maximal ℓ-ideal of $\mathrm{PL}(X)$ (or of $\mathrm{WPL}(X)$), then $\mathrm{Aff}(p)$ is a single point. In this case, $h(p)$ may be thought of as a real number. For a second example, let $p \subseteq \mathrm{PL}([-1,1])$ be the prime whose elements are those p.l. functions which vanish on some interval of the form $[0,\varepsilon)$ with $\varepsilon > 0$. Then $\mathrm{Aff}(p) = \mathbb{R}$. If $h \in \mathrm{PL}([-1,1])$, then h agrees with some affine linear function on a sufficiently small interval $[0,\varepsilon)$, and this is the function we call $h(p)$. For a third example, let $X = [-1,0) \cup (0,1] \subseteq [-1,1]$. Let $p_r \subseteq \mathrm{WPL}(X)$ (resp. $p_l \subseteq \mathrm{WPL}(X)$) be the maximal ℓ-ideal whose elements are those w.p.l. functions whose right-hand (resp. left-hand) limit at 0 is 0. Then $\mathrm{Aff}(p_r) = \{0\} = \mathrm{Aff}(p_l)$. If $h \in \mathrm{WPL}(X)$, then the right and left limits of h at 0 may differ. We have $h(p_r)$ = right-hand limit of h at 0 and $h(p_l)$ = left-hand limit of h at 0.

Our final example clearly illustrates some of the ideas which will be involved in the proof of theorem 3.1. Let $X = \{(x,y) \in [-1,1]^2 \mid |y| = x^2\}$ (c.f. example 2 in section 1). Let q_+ (resp., q_-) be the prime of $\mathrm{WPL}(X)$ whose elements are those w.p.l. functions which vanish on some set of the form $X \cap S_+$, where S_+ is a simplex with vertices $(0,0)$, $(\varepsilon,0)$ and (ε,δ), $\varepsilon,\delta > 0$ (resp., on some set of the form $X \cap S_-$, where S_- is a simplex with vertices $(0,0)$, $(\varepsilon,0)$ and $(\varepsilon,-\delta)$, $\varepsilon,\delta > 0$). Let p_+ (resp. p_-) be the smallest prime of $WPL(X)$ which properly contains q_+ (resp., q_-). Then $Aff(p_+) = Aff(p_-) = \mathbb{R} \times \{0\}$ (= the "x-axis"). Clearly, neither p_+ nor p_- is geometric. If $h:X \to \mathbb{R}$ is defined by

$$h(x,y) = \begin{cases} x & \text{if } y \geq 0 \\ 0 & \text{if } y \leq 0 \end{cases}$$

then (as mentioned in §2) h is w.p.l. but not strictly p.l. In this case, $h(p_+)(x,0) = x$ while $h(p_-)(x,0) = 0$.

The following lemma shows that the kind of behavior shown in the last example typifies w.p.l. functions which are not strictly p.l.

<u>Lemma 3.2.</u> Let $h \in WPL(X)$. Then $h \in PL(X)$ if and only if for every pair of primes $p_0,p_1 \subseteq WPL(X)$, if $p_0 \cap PL(X) = p_1 \cap PL(X)$ then $h(p_0) = h(p_1)$.

<u>Proof:</u> If $h \in PL(X)$ and p is any prime of $WPL(X)$, then $h(p) = h(p \cap PL(X))$. On the other hand, if $h \in WPL(X)$ satisfies the condition let W be the vector-lattice generated by $PL(X)$ together with h . Then every element of W also satisifes the condition so primes of $WPL(X)$ with the same intersection with $PL(X)$ are not separated by any element of W . Thus the primes of W correspond one-to-one to primes of $PL(X)$ so $W = PL(X)$, by the results of §2. □

We are now ready for the proof of 3.1. Given h weakly but not strictly p.l., pick a prime $p \subseteq PL(X)$ at which the condition of the lemma fails. (It may help the reader to keep the case in which p is maximal (i.e. a point of I^n) in mind.) As q varies over the primes of $WPL(X)$ such that $q \cap PL(X) = p$, there are at least two and at most finitely many distinct values assumed by $h(q)$. Call them

$b_1, \ldots, b_m$: Aff(p) $\to \mathbb{R}$. Each b_i is the restriction to Aff(p) of a linear function $a_i: \mathbb{R}^n \to \mathbb{R}$ appearing in a presentation of h and for any $i = 1, \ldots, m$ and any neighborhood of p , there are points x of X in that neighborhood at which $h(x) = a_i(x)$. Assume that $a_1 < a_2 < \ldots < a_m$ (mod p) . A sufficiently small neighborhood U of p will meet the following conditions:

i) $\forall \in U \cap X$, $h(x) = a_1(x)$ or $h(x) = a_2(x)$ or ... or $h(x) = a_m(x)$

ii) $\forall \in U \cap X$, $a_1(x) < a_2(x) < \ldots < a_m(x)$.

The data (i) and (ii) allow us to construct a splitting of X near p . Let $d_i = \frac{1}{2}(a_{i+1} + a_i)$. Then $a_1 < d_1 < a_2 < d_2 < \ldots < d_{m-1} < a_m$ on $U \cap X$. Thus $U \cap X$ is the disjoint union of the sets $\{x \in U \cap X | \ h(x) < d_1(x)\}$, $\{x \in U \cap X \mid d_1(x) < h(x) < d_2(x)\}$, ... , $\{x \in U \cap X \mid d_{m-1}(x) < h(x)\}$. That p is in the closure of each of these sets follows from the fact mentioned in the sentence introducing the a_i's.

For the converse, assume X splits inside U near p . Because of the nature of $\mathfrak{D}$, we may assume there are two closed p.l. sets Y_1 and Y_2 such that $Y_1 \cup Y_2 \supseteq U \cap X$ and $Y_1 \cap Y_2 \cap X \subseteq \text{Aff}(\mathfrak{q})$, where $\mathfrak{q}$ is the smallest prime of PL(X) containing p . (If p is maximal, Aff($\mathfrak{q}$) = $\emptyset$). Pick positive p.l. functions $f_1, f_2, g: X \to \mathbb{R}$ such that $f_1 = f_2$ (mod $\mathfrak{q}$), $f_1 < f_2 < g$ (mod p) and the support of g is contained in U . Define $h: X \to \mathbb{R}$ by $h = f_1 \wedge g$ on Y_1 and $h = f_2 \wedge g$ on $Y_2 \cap U$ and h = 0 outside U . Then h is weakly p.l. and not p.l. This completes the proof of the theorem.

As a final remark, we point out that the splitting condition mentioned in the theorem really is a geometric property of the embedding of X in $[-1,1]^n$. Essentially, this is because all primes of $PL([-1,1]^n)$ are geometric (Baker), and $\text{SpecPL}(X) \subseteq \text{SpecPL}([-1,1]^n)$. Thus, all the conditions in the theorem can be phrased in terms of filters of p.l. sets.

ACKNOWLEDGMENT

The author would like to thank J. R. Isbell for pointing out the relevance of Michael's work to the subject of this paper.

REFERENCES

1. M. Anderson and P. Conrad, Epicomplete ℓ-groups, Alg. Univ. 12(1981), 224-241.

2. K. A. Baker, Free vector lattices, Canad. J. Math. 20(1968) 58-66.

3. W. M. Beynon, Duality theorems for finitely generated vector lattice, Proc. London Math. Soc. (3)31(1975), 114-128.

4. L. Gillman and M. Jerison, Rings of continuous functions, Van Nostrand, Princeton, 1960.

5. E. Michael, Cuts, Acta Math. 111(1964), 1-36.

BANASCHEWSKI FUNCTIONS AND RING-EMBEDDINGS

by R.H. Redfield,
Hamilton College, Clinton, New York 13323, U.S.A.

1. Introduction

The examples of totally ordered abelian groups which are most easily constructed are the lexicographically ordered products of copies of the real numbers. If the unordered product is formed over a totally ordered index set, then the lexicographically ordered product consists of the functions in the unordered product with inversely well-ordered supports. These functions are added coordinatewise, and a function is greater than zero if its value at its maximal supporting element is greater than zero. The resulting structure is a totally ordered abelian group.

In 1907, Hahn [7] proved that every totally ordered abelian group can be embedded in a lexicographically ordered product of the kind described above. His proof was long and, over time, other methods were introduced which shortened it considerably (cf. [2]). One of these methods entailed the use of auxiliary functions which have come to be known eponymously as Banaschewski functions [1].

Examples of lattice-ordered, rather than totally ordered, abelian groups may be produced in much the same way. In this case, choose as an index set a partially ordered set for which the set of elements above any given element is totally ordered; again form the unordered product. In this case, the functions selected from the unordered product are those for which the set of supporting elements above any given supporting element is inversely well-ordered. The functions in the resulting set are added coordinatewise, and a function is greater than zero if all of its values at its maximal supporting elements are greater than zero. The resulting structure is a lattice-ordered abelian group [5].

In 1963, Conrad, Harvey and Holland [5] generalised Hahn's Theorem to the class of lattice-ordered abelian groups by showing that any such group can be embedded in a lattice-ordered product of the kind described

J. Martinez (ed.), Ordered Algebraic Structures, 247–255.

above. Included in the machinery they used were the ubiquitous functions of Banaschewski.

The validity of such theorems for groups raises hopes that similar theorems might be true for rings and fields. In fact, the result for fields, which is based on the power series fields over the real numbers introduced by Levi-Cività in 1893 (cf. [8](1)), has apparently been known at least since the 1950's(2).

The construction of Levi-Cività's power series fields is an extension of the construction of the totally ordered product described above. In this case, however, the index set must be a totally ordered abelian group so that the multiplication can be defined. The elements of the power series field are, as above, the functions in the unordered product with inversely well-ordered support, and also as above, addition is defined coordinatewise and the total order is defined by using the maximal elements of the supports. Since the index set is a group, its elements may also be multiplied together, and hence multiplication in the power series field may be defined by considering the functions as polynomials. That is, to find the value of a product of two functions at a given coordinate, sum the products of the values of the functions at pairs of coordinates whose product is the given coordinate. The resulting structure is a totally ordered field (cf. [10]).

The result analogous to Hahn's Theorem then asserts that every totally ordered field may be embedded in a totally ordered power series field. Although, as noted above(2), this theorem has apparently been known to field theorists for a long time, the first complete proof to appear in print, as far as the present author can determine, is the one by Conrad and Dauns in [4] (see also [11]). Conrad and Dauns also proved various results concerning lattice-ordered fields (see also [16]), but a general embedding theorem for such fields in the spirit of Hahn's Theorem has yet to be found.

The situation for rings is even murkier than that for fields. With the requirement that the index set be a semigroup or monoid rather than a group, the totally ordered power series fields of Levi-Cività become totally ordered power series rings. Whether every totally ordered ring can be embedded in such a power series ring is not known. However, the present author has characterised in [12] the totally ordered rings which can be embedded in totally ordered power series rings by means of embeddings analogous to those found in both Hahn's Theorem and its extensions to lattice-ordered groups and totally ordered fields. This characterisation de-

---

(1) The author thanks W.A.J. Luxemburg for bringing this reference to his attention (cf. [9]).

(2) At the meeting in Curaçao, Paul Conrad mentioned that field theorists had told him that they were aware of this result by the 1950's. Other mathematicians have also told the author this. In addition, see page 138 of [6].

pends on the totally ordered ring possessing a particular kind of Banaschewski function. Since there is a large class of embeddable rings which do not possess such functions[(1)] [13], this method of proof will not suffice to prove the most general embedding theorem for ordered rings.

However, a question which does remain is how far these techniques can be pushed to obtain embedding theorems for rings in general. The object of the following sections is to outline an answer to this question, an answer which in fact yields an embedding theorem for lattice-ordered rings. The details will appear separately [14].

2. General Power Series Rings

A ring $(T,+,\cdot)$ for which $(T,+)$ is a divisible abelian group we will call a **divisible ring**. For a divisible ring $(T,+,\cdot)$ and a semigroup $(\Delta,\cdot)$, we will construct a power series ring over Δ whose components are subgroups of T. If $(\Delta,\cdot)$ has a zero-element, denote it by ω.

Firstly, for $A,B \subseteq \Delta$, let $AB = \{\alpha\beta \mid \alpha \in A, \beta \in B\}$ and note that with respect to this multiplication $(2^{\Delta},\cdot,\subseteq)$ is a partially ordered semigroup. A subset X of 2^{Δ} is then a **supporting subset** if it satisfies the conditions:

(P_1) X contains all the atoms of the lattice $(2^{\Delta},\subseteq)$;

(P_2) X is an ideal of the lattice $(2^{\Delta},\subseteq)$;

(P_3) X is a subsemigroup of the semigroup $(2^{\Delta},\cdot)$;

(P_4) the set $\{(\alpha,\beta) \in A \times B \mid \alpha\beta = \delta\}$ is finite whenever $A,B \in X$ and $\omega \neq \delta \in \Delta$.

Choose divisible subgroups $\{C_\delta\}_{\delta \in \Delta}$ of T such that if $\gamma\delta \neq \omega$, then $C_\gamma C_\delta \subseteq C_{\gamma\delta}$, and form the product $\Pi_\Delta C_\delta$. This is clearly an abelian group with respect to coordinatewise addition: $(f+g)_\delta = f_\delta + g_\delta$. Let ${}_X\Pi_\Delta C_\delta$ denote the set of all elements of $\Pi_\Delta C_\delta$ which have their supports in X. By (P_2), $({}_X\Pi_\Delta C_\delta,+)$ is a subgroup of $(\Pi_\Delta C_\delta,+)$. If we define a multiplication on ${}_X\Pi_\Delta C_\delta$ by letting

(1) One such ring may be constructed by using the formalism described in §2 below. Let Q be the totally ordered commutative ring of rational numbers; let $\Delta = Z \overleftarrow{\times} Z$ be the lexicographic product of the integers with themselves; let X be the set of inversely well-ordered subsets of Δ; let $({}_X\Pi_\Delta Q,+,\cdot,\leq)$ be the totally ordered, commutative power series ring defined in §2; and let R be the subring of ${}_X\Pi_\Delta Q$ consisting of all $f \in {}_X\Pi_\Delta Q$ for which the set $\{n \mid f_{(m,n)} \neq 0 \text{ for some } m\}$ is finite. Choose $s \in R$ such that $s_{(0,0)} = 1 = s_{(0,-1)}$ and $s_{(m,n)} = 0$ otherwise, and define a new multiplication $\blacklozenge$ on R by letting $f \blacklozenge g = fgs$. Since $s > 0$, $(R,+,\blacklozenge,\leq)$ is also a totally ordered commutative ring. The function $\mu(x) = xs$ is a one-to-one, order-preserving ring-homomorphism of $(R,+,\blacklozenge,\leq)$ into $({}_X\Pi_\Delta Q,+,\cdot,\leq)$, but, according to [13], $(R,+,\blacklozenge,\leq)$ does not possess the desired Banaschewski function.

$$(fg)_\delta = \begin{cases} \sum_{\alpha\beta=\delta} f_\alpha g_\beta & \text{if } \delta \neq \omega \\ 0 & \text{if } \delta = \omega \end{cases}$$

then $(\mathrm{x}\Pi_\Delta C_\delta,+,\cdot)$ is a ring (cf. [3]) called a **general power series ring**.

If the subgroups on which a general power series ring is based are totally ordered, if the index set has an appropriate compatible order, and if the supporting subset consists of appropriately ordered sets, then the power series ring may be given a compatible lattice order in the following way.

Firstly, recall some terminology. A partially ordered set $(\Delta,\leq)$ is a **root system** if $[\delta,\infty)$ is totally ordered for all $\delta \in \Delta$; if $\Gamma \subseteq \Delta$, then Γ is **locally inversely well-ordered** if, for all $\gamma \in \Gamma$, $[\gamma,\infty) \cap \Gamma$ is inversely well-ordered. A partially ordered semigroup $(\Delta,\cdot,\leq)$ for which $(\Delta,\leq)$ is a root system is a **rooted semigroup**. If $(\Delta,\cdot,\leq)$ is a partially ordered semigroup, then 2^Δ contains a supporting subset of locally inversely well-ordered sets if and only if $(\Delta,\cdot,\leq)$ is a rooted semigroup.

With this terminology in mind, we suppose that $(\Delta,\cdot,\leq)$ is a rooted semigroup, that $(T,+,\cdot,\leq)$ is a divisible lattice-ordered ring, that $\{C_\delta\}_{\delta\in\Delta}$ is a collection of totally ordered divisible subgroups of T such that if $\gamma\delta \neq \omega$, then $C_\gamma C_\delta \subseteq C_{\gamma\delta}$, and that $X \subseteq 2^\Delta$ is a supporting subset consisting entirely of locally inversely well-ordered sets. (For example, X could be the collection of all finite subsets of Δ.) Define a relation $<$ on $\mathrm{x}\Pi_\Delta C_\delta$ by letting $f < g$ if and only if $0 < (f - g)_\mu$ for all maximal elements μ in the support of $f - g$. Then $(\mathrm{x}\Pi_\Delta C_\delta,+,\cdot,\leq)$ is a lattice-ordered ring(1) and, if $(\Delta,\cdot,\leq)$ is totally ordered, then $(\mathrm{x}\Pi_\Delta C_\delta,+,\cdot,\leq)$ is a totally ordered ring (cf. [3], [4], [7], [8], [10]).

3. Banaschewski Functions

Let $(T,+,\cdot)$ be a divisible ring and let $\mathbf{D}$ denote the set of divisible subgroups of T. The following terminology is adapted from that developed for lattice-ordered groups in [5]. A subset C of D is **full** if for all $0 \neq x \in T$, there exists $M \in C$ such that M is maximal in C without x;

(1) At the meeting in Curaçao, Paul Conrad asked for a general way of constructing lattice-ordered fields and John Dauns asked in a problem session for an example of a proper lattice-ordered division ring. The present author described to each of them a way of constructing archimedean lattice-ordered fields and division rings by "tipping over" totally ordered fields and division rings. It is an easy matter to use this technique and the lattice-ordered power series rings defined here to produce many nonarchimedean lattice-ordered fields and division rings in which 1 is greater than 0 and many others in which 1 is not greater than 0 (see [15]).

in this case, a **C-value** of x is one of the elements of C which are maximal in C without x. We let MC denote the set of all possible C-values. A **Banaschewski lattice** is then a full subset C of D such that

(L_1) C contains both $\{0\}$ and T;

(L_2) C is closed with respect to arbitrary intersection.

A Banaschewski lattice C is clearly a complete lattice with respect to set inclusion and furthermore every C-value M has a unique cover $M^\wedge$ in C. A subset S of C is then C-**full** if for all $0 \neq x \in T$, there exists a C-value M of x such that both M and its cover $M^\wedge$ are in S.

Now let $\tau: D \to D$, let $S_\tau = \{T\} \cup \{C \in D \mid \tau(C) \neq \{0\}\}$, and suppose that τ satisfies the following conditions:

(B_1) S_τ is C_τ-full for some Banaschewski lattice C_τ;

(B_2) if $M \subseteq N$ in S_τ, then $\tau(M) \supseteq \tau(N)$;

(B_3) for all $M \in S_\tau$, $T = M \oplus \tau(M)$.

For such a τ, let $V[\tau] = MC_\tau \cap S_\tau$ and for $M \in V[\tau]$ let $M^\dagger = M^\wedge \cap \tau(M)$. Let $V[\tau]^\circ = V[\tau] \cup \{T\}$, and define a binary operation $\Diamond$ on $V[\tau]^\circ$ as follows: If $M,N \in V[\tau]$ and K is unique in $V[\tau]$ such that $M^\dagger N^\dagger \subseteq K^\dagger$, then $M \Diamond N = K$; otherwise $M \Diamond N = T$. To ensure that τ behaves well with respect to multiplication, we make the following further assumption:

(B_4) if $M,N \in V[\tau]$ then $\tau(M)\tau(N) \subseteq \tau(M \Diamond N)$.

A function $\tau: D \to D$ satisfying conditions (B_1) to (B_4) is called a **Banaschewski function** (cf. [1], [5]). A **β-ring** $(T,+,\cdot,\tau)$ is then a divisible ring $(T,+,\cdot)$ together with a Banaschewski function $\tau: D \to D$ (cf. [12]).

If the index set of a general power series ring is a rooted semigroup, then the power series ring has a natural Banaschewski function defined on it in the following way.

Suppose that $(\Delta,\cdot,\leq)$ is a rooted semigroup, that $(T,+,\cdot)$ is a divisible ring, and that $X \subseteq 2^\Delta$ is a supporting subset consisting entirely of locally inversely well-ordered sets. Choose subgroups $\{C_\delta\}_{\delta\in\Delta}$ of T such that if $\gamma\delta \neq \omega$, then $C_\gamma C_\delta \subseteq C_{\gamma\delta}$ and if $\delta \neq \omega$, then $C_\delta \neq \{0\}$; form the general power series ring $({}_X\Pi_\Delta C_\delta,+,\cdot)$. For $\Gamma \subseteq \Delta$ and $\gamma \in \Delta$, let

$$\downarrow\Gamma = \{f \in {}_X\Pi_\Delta C_\delta \mid f_\gamma = 0 \text{ for all } \gamma \in \Gamma\},$$

$$\uparrow\Gamma = \{f \in {}_X\Pi_\Delta C_\delta \mid f_\delta = 0 \text{ for all } \delta \notin \Gamma\}.$$

We say that a subset Γ of Δ is a **ray** if $[\gamma,\infty) \subseteq \Gamma$ whenever $\gamma \in \Gamma$ and a **principal ray** if Γ is of the form $[\gamma,\infty)$ or (γ,∞) for some $\gamma \in \Delta$. We then define $\chi: D \to D$ by letting

$$\chi(D) = \begin{cases} \uparrow\Gamma & \text{if } D = \downarrow\Gamma \text{ for some principal ray } \Gamma \\ \{0\} & \text{otherwise.} \end{cases}$$

Then for $C_\chi = \{\downarrow\Gamma \mid \Gamma \text{ is a ray of } \Delta\}$, χ is a Banaschewski function and hence $({}_X\Pi_\Delta C_\delta,+,\cdot,\chi)$ is a β-ring.

Note that since any semigroup becomes a rooted semigroup when given the trivial order, the above construction applies to any unordered general power series ring. Note also that if $(T,+,\cdot,\leq)$ is a totally ordered ring, then with the order $\leq$ defined as in §2, $(x\Pi_{\Delta}C_{\delta},+,\cdot,\leq,\chi)$ is a lattice-ordered β-ring.

4. Ring-embeddings

The embedding theorems mentioned at the end of §1 use Banaschewski functions to determine embeddings of rings into power series rings. These theorems start with a divisible ring $(T,+,\cdot)$ and attempt to determine when it can be embedded in a general power series ring based on an indexed collection $\{C_{\delta}\}_{\delta\in\Delta}$ of its nonzero divisible subgroups. Since the ring T may come equipped with an order, it is also assumed that the index set Δ has a predetermined order $\leq$.

Forming a general power series ring from the subgroups $\{C_{\delta}\}_{\delta\in\Delta}$ requires that a compatible operation be defined on Δ. To define such an operation, firstly adjoin a new element ω to Δ: $\Delta^{\circ} = \Delta \cup \{\omega\}$ and $C_{\omega} = \{0\}$. On Δ°, define a binary operation $\Diamond$ as follows: If $\gamma,\delta \in \Delta$ and μ is unique in Δ such that $C_{\gamma}C_{\delta} \subseteq C_{\mu}$, then $\gamma \Diamond \delta = \mu$; otherwise $\gamma \Diamond \delta = \omega$. (Note that this operation $\Diamond$ is a general version of the similarly denoted operation $\Diamond$ defined on $V[\tau]^{\circ}$ in §3.) While it is possible that $(\Delta^{\circ},\Diamond)$ is not a semigroup, it frequently is, in which case we say that $\{C_{\delta}\}_{\delta\in\Delta}$ is **semigroup-generating**. Thus, whenever $\{C_{\delta}\}_{\delta\in\Delta}$ is semigroup-generating, we may form power series rings $(x\Pi_{\Delta^{\circ}}C_{\delta},+,\cdot)$, and whenever $(\Delta^{\circ},\Diamond,\leq)$ is a rooted semigroup, we may form β-rings $(x\Pi_{\Delta^{\circ}}C_{\delta},+,\cdot,\chi)$.

In summary, we begin with a partially ordered set $(\Delta,\leq)$, a divisible ring $(T,+,\cdot)$, and a collection $\{C_{\delta}\}_{\delta\in\Delta}$ of nonzero divisible subgroups of T. The collection $\{C_{\delta}\}_{\delta\in\Delta}$ is then a **Hahn-decomposition** of T if $(\Delta^{\circ},\Diamond,\leq)$ is a rooted semigroup for which there exists a supporting subset X of locally inversely well-ordered subsets of Δ° and a one-to-one ring-homomorphism $\Phi: T \to x\Pi_{\Delta^{\circ}}C_{\delta}$ such that

(H_1) for all principal rays Γ of Δ and all $x \in T$, $x = y + z$ for some $y \in \Phi^{-1}(\downarrow\Gamma)$ and $z \in \Phi^{-1}(\uparrow\Gamma)$;

(H_2) for all $x \in C_{\gamma}$, $\Phi(x)_{\delta} = \begin{cases} x & \text{if } \gamma = \delta \\ 0 & \text{otherwise.} \end{cases}$

The function Φ occurring in the above definition of a Hahn-decomposition is called a **Hahn-embedding**.

The theorems below assert that possessing a Hahn-decomposition is equivalent to being a β-ring. Roughly speaking, $(V[\tau]^{\circ},\Diamond,\subseteq)$ corresponds to $(\Delta^{\circ},\Diamond,\leq)$ and $\{C_{\delta}\}_{\delta\in\Delta}$ to $\{M^{\dagger}\}_{M\in V[\tau]}$. The component of an element in T at $M^{\dagger}$ is determined in the following way. If $M \in V[\tau]$ for a β-ring $(T,+,\cdot,\tau)$, then $T = M^{\dagger} \oplus M \oplus \tau(M^{\wedge})$, and hence for all $x \in T$ and $M \in V[\tau]$, there exist unique elements $x \| M \in M^{\dagger}$, $x \downarrow M \in M$, and $x \uparrow M \in \tau(M^{\wedge})$ such that $x = x\|M + x\downarrow M + x\uparrow M$. To denote the support of x with respect to these components, we use $S(x) = \{M \in V[\tau] \mid x\|M \neq 0\}$.

The general embedding theorem is the following.

Theorem 1. *Let $(\Delta,\leq)$ be a partially ordered set, let $(T,+,\cdot)$ be a divisible ring, and let $\{C_\delta\}_{\delta \in \Delta}$ be a semigroup-generating collection of nonzero divisible subgroups of T. The following statements are equivalent:*

(I) *$\{C_\delta\}_{\delta\in\Delta}$ is a Hahn-decomposition of T.*

(II) *There exists a function $\tau: D \to D$ such that $(T,+,\cdot,\tau)$ is a β-ring for which the following hold:*
  - (a) *there is an order-isomorphism $V: (\Delta,\leq) \to (V[\tau],\subseteq)$ such that for all $\delta \in \Delta$, $V(\delta)^{\dagger} = C_\delta$;*
  - (b) *for all $M,N \in V[\tau]$, either $M \supseteq N$ or $M^{\dagger} \subseteq N$;*
  - (c) *for all $x,y \in T$, $S(xy) \subseteq S(x) \lozenge S(y)$;*
  - (d) *for all $x_1, \ldots, x_n \in T$, $(S(x_1) \lozenge \cdots \lozenge S(x_n),\subseteq)$ is locally inversely well-ordered;*
  - (e) *the set $\{ (K,L) \in S(x) \times S(y) \mid K \lozenge L = M \}$ is finite for all $x,y \in T$ and $M \in V[\tau]$.*

In the unordered situation, Δ is given the trivial order and Theorem 1 becomes the following.

Theorem 2. *Let $(T,+,\cdot)$ be a divisible ring, and let $\{C_\delta\}_{\delta\in\Delta}$ be semigroup-generating collection of nonzero divisible subgroups of T. The following statements are equivalent:*

(I) *$\{C_\delta\}_{\delta\in\Delta}$ is a Hahn-decomposition of T with respect to the trivial order on Δ.*

(II) *There exists a supporting subset X of $2^{\Delta^{\circ}}$ and a one-to-one ring-homomorphism $\Phi: T \to {}_X\Pi_{\Delta^{\circ}} C_\delta$ such that*

$$(\mathrm{H}_2) \text{ for all } x \in C_\gamma,\ \Phi(x)_\delta = \begin{cases} x & \text{if } \gamma = \delta \\ 0 & \text{otherwise.} \end{cases}$$

(III) *There exists a function $\tau: D \to D$ such that $(T,+,\cdot,\tau)$ is a β-ring for which the following hold:*
  - (a) *there is a bijection $V: \Delta \to V[\tau]$ such that for all $\delta \in \Delta$, $V(\delta)^{\dagger} = C_\delta$;*
  - (b) *for all distinct $M,N \in V[\tau]$, $M^{\dagger} \subseteq N$;*
  - (c) *for all $x,y \in T$, $S(xy) \subseteq S(x) \lozenge S(y)$;*
  - (d) *the set $\{ (K,L) \in S(x) \times S(y) \mid K \lozenge L = M \}$ is finite for all $x,y \in T$ and $M \in V[\tau]$.*

If $(T,+,\cdot,\leq)$ is a lattice-ordered ring and each C_δ is totally ordered, then the natural order to place on Δ is the following: $\alpha \circledless \beta$ if and only if $x < y$ for all $x \in C_\alpha$ and all $0 < y \in C_\beta$. If the Hahn-embedding Φ is to be a lattice-homomorphism, then the elements of $V[\tau]$ must also be suitably restricted. Recall that a **prime** subgroup of a lattice-ordered group $(T,+,\leq)$ is a convex lattice-subgroup P for which $0 \leq a \wedge b \in P$ implies that $a \in P$ or $b \in P$. In view of [5], the natural restriction on $V[\tau]$ is that its elements be prime subgroups. Specifically, we say that $(T,+,\cdot,\leq,\tau)$ is a **prime β-ring** if $(T,+,\cdot,\leq)$ is a lattice-ordered ring and $(T,+,\cdot,\tau)$ is a β-ring with an accompanying Banaschewski lattice C_τ for which every element of $V[\tau]$ is prime. In this case, Theorem 1 becomes the following.

Theorem 3. *Let $(T,+,\cdot,\leq)$ be a lattice-ordered divisible ring, and let $\{C_\delta\}_{\delta\in\Delta}$ be a collection of totally ordered, nonzero, divisible subgroups of T such that $(\Delta^\circ,\Diamond,\circledS)$ is a partially ordered semigroup. Then the following statements are equivalent:*

(I) *$\{C_\delta\}_{\delta\in\Delta}$ is a Hahn-decomposition of T whose Hahn-embedding is a lattice-homomorphism.*

(II) *There exists a function $\tau\colon D \to D$ such that $(T,+,\cdot,\tau)$ is a prime β-ring for which the following hold:*
- (a) *there is an order-isomorphism $V\colon (\Delta,\leq) \to (V[\tau],\subseteq)$ such that for all $\delta \in \Delta$, $V(\delta)^\dagger = C_\delta$;*
- (b) *for all $M,N \in V[\tau]$, either $M \supseteq N$ or $M^\dagger \subseteq N$;*
- (c) *for all $x,y \in T$, $S(xy) \subseteq S(x) \Diamond S(y)$;*
- (d) *for all $x_1, \ldots, x_n \in T$, $(S(x_1) \Diamond \cdots \Diamond S(x_n),\subseteq)$ is locally inversely well-ordered;*
- (e) *the set $\{ (K,L) \in S(x) \times S(y) \mid K \Diamond L = M \}$ is finite for all $x,y \in T$ and $M \in V[\tau]$;*
- (f) *for all $K \in V[\tau]$ and $0 \neq z \in \tau(K)$, there exists $M \in S(z)$ containing K.*

The most restricted situation occurs when $(T,+,\cdot,\leq)$ is a totally ordered ring without zero-divisors. In this case, if we let $(A[\tau],\Diamond,\leq)$ denote the totally ordered semigroup of archimedean classes of T (cf. page 128 of [6]), then Theorem 1 becomes the following (cf. [12]).

Theorem 4. *Let $(T,+,\cdot,\leq)$ be a totally ordered divisible ring without zero-divisors, let $\{C_\delta\}_{\delta\in\Delta}$ be a disjoint collection of nonzero, divisible, archimedean subgroups of T, and let X be the set of inversely well-ordered subsets of $A[\tau]$. Then X is a supporting subset of $2^{A[\tau]}$, and the following statements are equivalent.*

(I) *$\{C_\delta\}_{\delta\in\Delta}$ is a Hahn-decomposition of T whose Hahn-embedding Φ satisfies the condition*

$x \leq y$ *if and only if* $\Phi(x) \leq \Phi(y)$.

(II) *There exists a function* $\tau: D \to D$ *and an order-isomorphism* $V: (A[\tau],\leq) \to (V[\tau],\subseteq)$ *such that* $(T,+,\cdot,\leq,\tau)$ *is a prime* β*-ring and for all* $\alpha \in A[\tau]$, $V(\alpha)^{\dagger} = C_{\alpha}$.

The simplicity of Theorem 4 as opposed to Theorem 3 may raise the hope that some of the conditions of Theorem 3 are redundant. However, since there exist examples showing that the conditions (a) through (f) of Theorem 3 are independent of each other [14], Theorem 3 would seem to be the best possible general embedding theorem whose proof relies on Banaschewski functions.

REFERENCES

1. B. Banaschewski,*Totalgeordnete Moduln*, Arch. Math. 7 (1956), 430-440.
2. A.H. Clifford, *Note on Hahn's theorem on ordered Abelian groups*, Proc. Amer. Math. Soc. 5 (1954), 860-863.
3. P. Conrad, *Generalized semigroup rings*, J. Indian Math. Soc. 21 (1957), 73-95.
4. P. Conrad, J. Dauns, *An embedding theorem for lattice-ordered fields*, Pacific J. Math. 30 (1969), 385-398.
5. P. Conrad, J. Harvey, C. Holland, *The Hahn embedding theorem for lattice-ordered groups*, Trans. Amer. Math. Soc. 108 (1963), 143-169.
6. L. Fuchs, *Partially Ordered Algebraic Systems*, Pergamon Press, Oxford 1963.
7. H. Hahn, *Über die nichtarchimedischen Grossensysteme*, S.-B. Wiener Akad. Math.-Nat. Klasse Abt. IIa 116 (1907), 601-653.
8. T. Levi-Cività, *Sugli infiniti ed infinitesimi attuali quali elementi analitia*, Accademia Nazionali Sci., Lett. Arti, ser 7a, 4 (1892-1893), 1765-1815.
9. W.A.J. Luxemburg, *On a class of valuation fields introduced by A. Robinson*, Israel J. Math. 25 (1976), 189-201.
10. B.H. Neumann, *On ordered division rings*, Trans. Amer. Math. Soc. 66 (1949), 202-252.
11. S. Priess-Crampe, *Zum Hahnschen Einbettungssatz für angeordnete Körper*, Arch. Math. 24 (1973), 607-614.
12. R.H. Redfield, *Embeddings into power series rings*, manuscripta math. 56 (1986), 247-268.
13. R.H. Redfield, *Non-embeddable o-rings*, Comm. in Algebra 17(1) (1989), 59-71.
14. R.H. Redfield, *Representations of rings via Banaschewski functions*, (to appear).
15. R.H. Redfield, *Constructing lattice-ordered fields and division rings*, (to appear).
16. S. Steinberg, *An embedding theorem for commutative lattice-ordered domains*, Proc. Amer. Math. Soc. 31 (1972), 409-416.

ON ORDERPOTENT RINGS

Piotr Wojciechowski

Bowling Green State University

In [1] has been given the definition and some basic properties of orderpotent rings, at that time called ℓ-nilpotent rings (see the definition below). The main theorem concerned the structure of orderpotent algebras over a totally ordered ring. Presently we restate the theorem along with a much better proof. We also give the answers to the two questions asked at the end of that paper.

1. **Definition.** A partially ordered ring $\mathcal{R}$ is said to be *n-orderpotent* (or *orderpotent of degree n*), if for each n elements $a_1,\dots,a_n$ of $\mathcal{R}$:

$$\text{either} \quad a_1\cdot \dots \cdot a_n \geq 0 \quad \text{or} \quad a_1\cdot \dots \cdot a_n \leq 0,$$

that is each n-product of elements of $\mathcal{R}$ is related to zero.

Of course, every totally ordered ring is orderpotent; also any n-nilpotent p.o. ring is n-orderpotent; orderpotency is preserved by order-preserving homomorphisms and under taking p.o.-subrings. The ring of all upper trianglar $n\times n$ matrices over a totally ordered ring, with 0's on the main diagonal is $(n$-$1)$-orderpotent. Also the ring of zero-constant-term polynomials over an n-orderpotent ring becomes n-orderpotent, if we impose the following order:

$$a_1x + \dots + a_mx^m > 0 \text{ if } m > n \text{ and } a_m > 0$$

and

$$a_1x + \dots + a_mx^m \geq 0 \text{ if } m \leq n \text{ and } a_i \geq 0 \text{ for all } i = 1,\dots,m .$$

Any orderpotent ring with identity is totally ordered.

J. Martinez (ed.), Ordered Algebraic Structures, 257–262.

It may be easily seen that if a p.o. ring is a cardinal product of two p.o. rings, and at the same time is n-orderpotent, then at least one of the factors is n-nilpotent.

It follows that in the class of f-rings every n-orderpotent ring satisfies the condition:

(1) $$a_+^n = 0 \quad \text{or} \quad a_-^n = 0 \qquad \text{for any } a \text{ in the ring.}$$

2. The question is: does (1) imply that the given f-ring is n-orderpotent? The negative answer is provided by the following

Example 1. Let $\mathcal{R}_0$ be a p.o. zero-ring which is not totally ordered, e.g.

$$\mathcal{R}_0 = \mathbf{Z}_0 \oplus \mathbf{Z}_0$$

where $\mathbf{Z}_0$ denotes the zero-ring, whose additive group is that of the integers.Then let us consider $\mathcal{A} = \mathbf{Z} \overrightarrow{\times} \mathcal{R}_0$, that is the lexicographic product of $\mathbf{Z}$ (the usual totally ordered ring of the integers) and $\mathcal{R}_0$, and impose the usual addition on it, while the multiplication is given by:

$$(n,r)(m,s) = (nm, ns + mr)$$

Thus , this is nothing but the embedding of $\mathcal{R}_0$ into a p.o. ring with identity in the standard way. Now $\mathcal{A}$ is not totally ordered since $\mathcal{R}_0$ was not either. Since $\mathcal{A}$ has the identity element (1,(0,0)), it cannot be orderpotent. But $\mathcal{A}$ satisfies (1) with the power $n = 2$ for:

if $m \neq 0$ then $a = (m,s)$ is related to zero and either a_+ or a_- is zero;

if $m = 0$ then $a = (0,s)$ and $a^2 = 0$ so $a_+^2 = 0$ and $a_-^2 = 0$.

It remains to show that $\mathcal{A}$ is an f-ring. Indeed, take $a, b, c \in \mathcal{A}$ such that

$$a \perp b \ , c > 0.$$

Then $a = (0,\alpha)$, $b = (0,\beta)$ with $\alpha \perp \beta$. Let $c = (n,\gamma)$. If n is not zero, we have

$$ca \wedge b = (n,\gamma)(0,\alpha) \wedge (0,\beta) = (0,n\alpha) \wedge (0,\beta) = 0\ .$$

If $n = 0$, $ca \wedge b = (0,\gamma)(0,\alpha) \wedge (0,\beta) = (0,(0,0)) \wedge (0,\beta) = 0$. Since $\mathcal{A}$ is a commutative ring, $ac \wedge b = 0$ as well; so $\mathcal{A}$ is an f-ring.

3. Now we consider the canonically ordered algebras over a totally ordered field, i.e. we will refer to an algebra $\mathcal{A}$ as a *canonically ordered algebra* if

(*i*) for a given basis $\{e_i\}$ of $\mathcal{A}$ an element a in $\mathcal{A}$ satisfies: $a \geq 0$ if and only if all its coefficients in the given basis are non-negative (in this case we deal with the *canonically ordered vector space*), and

(*ii*) all "structure constants" are non-negative, i.e. $e_i e_j \geq 0$ for all i, j.

In order to prove the main theorem of this section, we will need the following lemmas.

Lemma 1. *Let V be a vector space with dim $V > 1$ over a totally ordered field; let "$\circ$" denote an inner product on V, and consider two vectors u and v from V. If for any vector w from V,*

$$(w \circ u)(w \circ v) \geq 0,$$

then u and v are linearly dependent.

This lemma is easily proved by assuming that u and v are independent and obtaining the contradiction in the two dimensional subspace generated by u and v (it is done in detail in [1]).//

Lemma 2. *Let $\mathcal{A}$ be a canonically ordered vector space over a totally ordered field* **F**, *and let P and Q belong to $\mathcal{A}$. Then if for every two scalars α and β,*

(2) $$\alpha P + \beta Q \geq 0 \text{ or } \alpha P + \beta Q \leq 0,$$

then the vectors P and Q are dependent.

Proof. Let $\{e_i\}$ be the canonical basis for $\mathcal{A}$. Then there is a positive integer m and the scalars $p_1, p_2, \dots, p_m; q_1, q_2, \dots, q_m$ such that

$$P = \sum_{t=1}^{m} p_t e_t, \quad Q = \sum_{t=1}^{m} q_t e_t,$$

so
$$\alpha P + \beta Q = \sum_{t=1}^{m} (\alpha p_t + \beta q_t) e_t.$$

By (2) and the canonical order of $\mathcal{A}$ it follows that for every $1 \leq t, s \leq m$, $\alpha p_t + \beta q_t$ and $\alpha p_s + \beta q_s$ have the same sign in **F**. We can express this fact by saying: in the **F**-vector space **F**$\oplus$**F** with the usual inner product, the vectors $[p_t, q_t]$ and $[p_s, q_s]$ satisfy the assumption of Lemma 1, therefore they are dependent. This means that the matrix

$$\begin{bmatrix} p_1 & p_2 & \cdots & p_m \\ q_1 & q_2 & \cdots & q_m \end{bmatrix}$$

has rank one. So the row vectors $(p_1,p_2,\dots,p_m)$ and $(q_1,q_2,\dots,q_m)$ are dependent, which yields dependence of P and Q. //

Theorem. *A canonically ordered algebra* $\mathcal{A}$ *is* n*-orderpotent if and only if* $\mathcal{A}^n$ *(subalgebra generated by all* n*-products of* $\mathcal{A}$*) is totally ordered.*

Proof. Part "$\Leftarrow$" is obvious for any orderpotent ring. To prove the "$\Rightarrow$" part, we show that $\mathcal{A}^n$ is one-dimensional subalgebra of $\mathcal{A}$.

Let $\mathcal{A}$ be n-orderpotent and let $\{e_i\}$ be its canonical basis. It is sufficient to show that any two n-products of *basic* elements are linearly dependent vectors. Let us take two such products:

(3) $$P = e_{i_1} \dots e_{i_n} \quad\text{and}\quad Q = e_{j_1} \dots e_{j_n}$$

and to avoid triviality suppose $P \neq 0$ and $Q \neq 0$.

The number k of true inequalities $e_{i_l} \neq e_{j_l}$ may vary from 0 to n, $l = 1,\dots,n$. If $k = 0$, then for all $l = 1,\dots,n$ $e_{i_l} = e_{j_l}$ and the vectors (3) are just equal, so suppose that $k > 0$.

We use induction on k to prove that the vectors (3) are dependent. Let $k = 1$ and let l be the index such that $e_{i_l} \neq e_{j_l}$. Then for any two scalars α and β,

$$\alpha P + \beta Q = e_{i_1} \dots e_{i_{l-1}}(\alpha e_{i_l} + \beta e_{j_l})e_{i_{l+1}} \dots e_{i_n}.$$

For each α and β, this element, being a product of n elements in an n-orderpotent algebra, is related to zero. Therefore, by Lemma 2, P and Q are dependent. Now suppose that the vectors (3) are dependent whenever $k < K$, and let $k = K$. If there is a non-zero vector $e_{s_1} \dots e_{s_n}$ for which the number k_1 of true inequalities $e_{s_l} \neq e_{i_l}$ is less than K, and the number k_2 of true inequalities $e_{s_l} \neq e_{j_l}$ is less than K, then this "between" vector is, by induction hypothesis, dependent of both vectors (3), therefore the vectors (3) are dependent too. Now let us suppose that every such "between" vector is the zero-vector, i.e. if for $e_{s_1} \dots e_{s_n}$ the k_1 and k_2 defined above satisfy $k_1 < K$ and $k_2 < K$, then $e_{s_1} \dots e_{s_n} = 0$. Let us take two arbitrary scalars α and β and consider the vector $v_{\alpha\beta} \in \mathcal{A}$ defined as follows: $v_{\alpha\beta} = f_1 \dots f_n$ where

$$f_l = \begin{cases} e_{i_l} & \text{if } e_{i_l} = e_{j_l} \\ \alpha e_{i_l} + \beta e_{j_l} & \text{if } l \text{ is the first index with } e_{i_l} \neq e_{j_l} \\ e_{i_l} + e_{j_l} & \text{otherwise} \end{cases}$$

Distributing, we may write $v_{\alpha\beta}$ as a sum of n-products of e_i's or e_j's with the coefficients α or β. But each such product is a "between" vector with respect to P and Q except for the vectors (3) themselves. According to our assumption that all those "between" vectors are zero, we reduce $v_{\alpha\beta}$ to

$$v_{\alpha\beta} = \alpha e_{i_1} \dots e_{i_n} + \beta e_{j_1} \dots e_{j_n} = \alpha P + \beta Q .$$

Since $\mathcal{A}$ is n-orderpotent, $v_{\alpha\beta}$ must be related to zero for any α and β. So again Lemma 2 gives us the required dependence of P and Q.//

4. We have shown that n-orderpotent canonically ordered algebras are just those whose n-power subalgebra is totally ordered. A natural question is to be asked about the general case: if $\mathcal{R}$ is a p.o. ring, then does the fact that $\mathcal{R}$ is n-orderpotent imply the total order of $\mathcal{R}^n$? It turns out to be not true as it is shown in the following

Example 2. Let us consider a free semigroup $\mathcal{S}$ on two generators a and b and for $s \in \mathcal{S}$, let $\ell(s)$ denote the length of s , that is , the number occurences of a and b in the unique expression of s as a product of the generators. Partially order it by letting, for s_1 and s_2 from $\mathcal{S}$:

$$s_1 < s_2 \text{ if and only if } \ell(s_1) \geq 2 \text{ and}$$

$$\text{either } \ell(s_1) < \ell(s_2)$$

or $\ell(s_1) = \ell(s_2)$ and s_1 preceeds s_2 in the lexicographic order, with the exception that

$$ab \parallel ba .$$

The generators are not non-trivially related to any element of $\mathcal{S}$.

$\mathcal{S}$ may be pictured by means of the diagram:

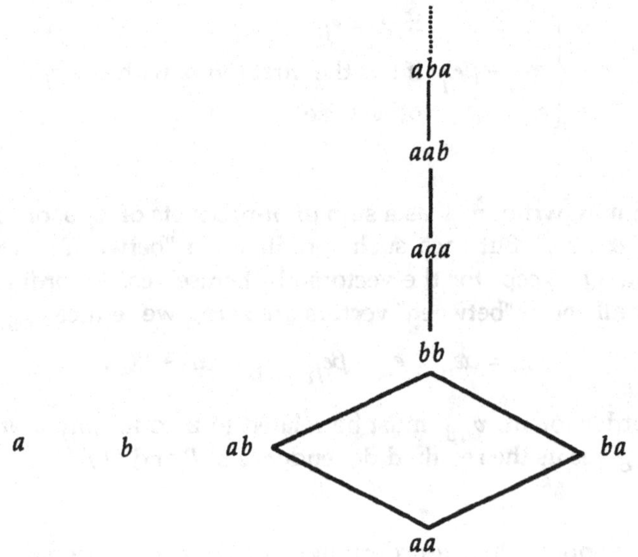

Now consider the semigroup ring $\mathcal{RS}$, where $\mathcal{R}$ is any totally ordered ring without zero divizors. Partially order it by letting p in $\mathcal{RS}$ be bigger than zero if the coefficients of the maximal elements of the support of p are bigger than zero. If we multiply any two elements x and y of $\mathcal{RS}$, we will always get exactly one maximal element of the support of xy. So $\mathcal{RS}$ is 2-orderpotent. But $(\mathcal{RS})^2$ is not totally ordered, for if $r \neq 0$, rab and rba are not related.

Reference

[1] Piotr Wojciechowski, *The concept of ℓ-nilpotent rings,* Algebra and Order. Proc. First Int. Symp. Ordered Algebraic Structures, Luminy-Marseilles, 1984.

A Sampling of Open Problems

> The town of Bowling Green is situated near the center of what was once known as The Great Black Swamp of Northwest Ohio. As an impediment to travel, it had a significant influence on the early history of the United States. Until the nineteenth century, it was an unsettled region dominated by "water, wildlife, immense vegetation, and miasmic vapors," to quote a history of the time. Then it was drained and cut, and the flat land with its black soil has since become one of the richest farming areas on earth. (From a presentation prepared by W. Charles Holland for the Special Session on Ordered Algebraic Structures held at the annual meeting of the American Mathematical Society in January, 1988.)

The Black Swamp Problem Book came into being in May of 1978. Over the years it has become the place to record, by hand, open problems which are judged to be interesting or significant by those who pose them. There are upwards of fifty problems recorded in the book; some of these have remained unsolved since the creation of the Black Swamp Problem Book.

Here we collect some problems from that record, as well as a number of others which were posed during the conference in Curaçao. As has already been observed, three of the articles in this volume represent solutions to problems presented during the conference. This account relies on the transcription of the Black Swamp Problem Book already mentioned above, as well as on a distillate of it which Holland prepared for Curaçao. I have added contributions collected by Conrad, Henriksen, Luxemburg and Prestel. A special thanks goes to these men for chairing problem sessions and rounding up the contributions.

The person whose name appears with the problem cited is taken to be either the proposer (at the conference) or else the author, according to the Black Swamp Problem Book. My apologies in advance for any mistakes in attribution.

J. Martinez (ed.), Ordered Algebraic Structures, 263–274.

The first two are among the oldest and toughest open problems about ordered groups.

1. Does every totallyordered group admit embedding into a totally ordered divisible group? (B.H. Neumann, ca. 1950) There has been very little progress on this problem since it was first posed.

2. Does every torsion-free abelian group admit an archimedean lattice order? (Elliot Weinberg, ca. 1962) I have, on at least two occasions, given false proofs for the affirmative. After periodic conversations with Laszlo Fuchs and Manfred Dugas I am now convinced that the answer is no, and that to find a counter-example one will need to employ some kind of fancy set-theoretic fireworks.

Ordered Groups

3. Can every finitely generated recursively presented ℓ-group be ℓ-embedded in a finitely presented ℓ-group? (Glass, 1972)

4. Is the word problem solvable for free products in the class of abelian ℓ-groups? Same question for other classes of ℓ-groups.

5. Let K be a class of ℓ-groups which is closed under ℓ-homomorphic images, convex ℓ-subgroups and free products. Is K a torsion class? (Powell & Tsinakis, 1980)

6. Does the class of vector lattices form a torsion class? (Claimed by Martinez, ca. 1975; Conrad first observed that the issue was by no means clear, one way or the other.)

7. Are any two scalar multiplications on a vector lattice connected by an ℓ-automorphism? (Conrad, 1982; The Hahn groups have this property.)

8. Suppose that on an ℓ-group G any two compatible group operations are connected by an ℓ-automorphism of the underlying lattice. Is G abelian? (Kopytov, 1984)

9. Does there exist a finitely generated simple right orderable group? (Rhemtulla, 1980) Same question, replacing "simple" by "perfect".

10. Does every right orderable polycyclic group possess a solvable series with torsion-free abelian factors? (Rhemtulla, 1980)

11. Is there a non-abelian variety of ℓ-groups which satisfies the amalgamation property? (Powell & Tsinakis, 1982; see Powell's contribution to this volume.)

12. For which varieties of ℓ-groups is the group of integers in the amalgamation base? (Powell & Tsinakis, 1982)

13. Is $\mathbf{Z} + \mathbf{Z}$ an amalgamation base for the class of all ℓ-groups? What about the lexicographic product of two copies of the integers? (K. Pierce, ?)

14. Is the variety of weakly abelian ℓ-groups the supremum (in the lattice of all varieties of ℓ-groups) of the varieties N_k, where N_k stands for the variety of nilpotent ℓ-groups of class not exceeding k? (Kopytov, 1984)

15. Does there exist any proper variety of ℓ-groups whose elements are recognizable from the underlying lattice? (Darnel, 1984)

16. Does every ℓ-group with a unique addition have to be archimedean? (Darnel, 1984)

17. If an abelian ℓ-group can be represented as a subdirect product of integers, does it admit a unique addition? (Conrad & Darnel, 1984) The answer is affirmative for many classes of ℓ-groups of this kind; among these are the class of Specker groups, the subdirect products of integers with a basis and the free abelian ℓ-groups; see Conrad & Darnel; paper to appear in the Transactions of the American Mathematical Society.

18. Suppose that G is an arbitrary ℓ-group. Min(G) stands for the space of minimal prime subgroups, endowed withthe hull-kernel topology. It is well known that this is a Hausdorff, 0-dimensional space. Conrad & Martinez have recently proved that Min(G) is compact if and only if G is complemented; that is, when $\mathbf{Pr}(G)$, the sublattice of principal polars, is a subalgebra of $\mathbf{P}(G)$, the boolean algebra of all polars of G. (This result extends similar ones for commutative rings and for vector lattices, which were in the literature for quite some time.)

The concern here is over complementations; that is, over the adjunction of elements to an ℓ-group so as to make it complemented, while at the same time Min(G) is compactified. We shall refrain from going into the technicalities of formally defining a complementation, and refer the reader to work of Conrad & Martinez, which is to appear soon.

We call a convex ℓ-subgroup A of the ℓ-group G a *z-subgroup* if for each $a \in A$ the entire principal polar generated by a lies in A. Consider then the following three conditions on the ℓ-group G:

(a) Every z-subgroup is an intersection of minimal primes. (Note: it is well known that an intersection of minimal primes is a z-subgroup.)

(b) G is locally complemented; (meaning that each principal convex ℓ-subgroup of G is complemented.)

(c) G admits a complementation.

Conrad & Martinez have shown that (a) is equivalent to (b), while (c) implies the other two. It is not known whether (a) implies (c). The implication is valid for finite-valued ℓ-groups; Martinez has, indeed, given a number of ℓ-group theoretic conditions for finite-valued ℓ-groups which are equivalent to the three conditions mentioned here. For *projectable* ℓ-groups, a great deal is known if one insists that G be embedded in a complementation as a convex ℓ-subgroup. However, even for projectable ℓ-groups it is not known whether the missing implication holds; there are examples of projectable ℓ-groups G which admit complementations, but none which contain G as a convex ℓ-subgroup.

Ordered Rings and Algebras

19. Let L be an archimedean Riesz space, and let $Z(L)$ be the center, of L; that is, $Z(L)$ is the ideal generated by the identity operator of L in the Riesz algebra $L_b(L)$ of all order bounded linear transformations of L.

Concerning $Z(L)$ the following seem to be open:

(a) What is the relationship, if any, between the structure space of an archimedean Riesz space and the structure space of $Z(L)$?

(b) Characterize the archimedean Riesz spaces for which the centers are finite dimensional.

(c) Which archimedean f-algebras have finite dimensional centers? (Luxemburg)

20. If G is a Riesz space and I is an ℓ-ideal, then G/I is archimedean if and only if I is relatively uniformly closed. What can be said when G is an ℓ-group? A partial answer may be obtained by examining the divisible hull of G. (Luxemburg)

21. If A is an archimedean Riesz algebra, then can the Dedekind-MacNeille completion of A as a Riesz space be given a Riesz algebra structure which induces the given one on A? (Huijsmans)

22. Let A be an archimedean f-algebra with identity 1. Let Specm-$\ell(A)$ be the space of maximal ring and ℓ-ideals of A (with the hull-kernel topology.) Specm$\mathbf{R}$-$\ell(A)$ is the subspace consisting of those ideals for which the factor is a copy of $\mathbf{R}$, the real numbers. For $A = C(X)$, where X is a Tychonoff space, it is well known that Specm-$\ell(A)$ is the Stone-Čech compactification of X; Specm$\mathbf{R}$-$\ell(A)$ is the real compactification.

In general, not much is known about Specm$\mathbf{R}$-$\ell(A)$. For example, under what conditions is it empty? dense in Specm-$\ell(A)$? equal to Specm-$\ell(A)$? If A is the algebra of all the real, Lebesgue-measurable functions modulo the null functions, then Specm-$\ell(A)$ is the Stone space of the Lebesgue measure algebra, but Specm$\mathbf{R}$-$\ell(A)$ is empty. (Luxemburg)

23. For each $n \geq 1$, in how many "different" ways can $\mathbf{R}^n$ be made into a Riesz algebra? (Huijsmans)

24. The celebrated Sikorski Extension Theorem for boolean algebras has been extended to Riesz spaces and Riesz homomorphisms; see Luxemburg & Schep, Proc. Royal Acad. of Sciences Amsterdam (A) 82 (1979), 145-54. The relative strengths of these theorems, as they relate to the Axiom of Choice (AC) in Zermelo-Fraenkel set theory (ZF) seem to be open. Obviously, AC implies both of the extension theorems, and the homomorphism extension theorem for Riesz spaces implies the Sikorski Extension Theorem. (See Gluschankof's article in this volume, as well as the papers by Luxemburg, Fund. Math. 55 (1964, 239-47 and Bell, Jour. of Symbolic Logic 48 (1983), 841-45.) (Luxemburg)

25. In the sense of nonstandard analysis, enlargements of ℓ-groups are again ℓ-groups. What structural properties do these enlargements possess? (See D. Cozart & L. Moore, Jr., Duke Math. Jour. 41 (1974), 263-275.) (Luxemburg)

26. Pierce & Birkhoff conjectured that every piecewise polynomial on $\mathbf{R}^n$ (which is, by definition, continuous and semi-algebraic) is a supremum of infima of finitely many polynomials. This is known for $n = 1$ and $n = 2$, but it is open for $n \geq 3$. If the conjecture is true for $n = 3$, then for any f, P and Q satisfying (∗) below, there is a g satisfying (∗∗). It is not known if such a g can always be found. A proof that it can would add significantly to our understanding of this conjecture. An example of f, P and Q for which no g exists would, of course, lay the whole issue to rest.

(∗) (a) $f \in \mathbf{R}[x_1, x_2, x_3]$ with f irreducible and $f(0,0,0) = 0$.

(b) P, $Q : \mathbf{R} \longrightarrow \mathbf{R}^3$ are rational curves so that $P(0) = Q(0) = (0,0,0)$.

(c) For small $\epsilon > 0$, $P((0,\epsilon))$ and $Q((0,\epsilon))$ are in different connected components of the set of points in $\mathbf{R}^3$ at which f is strictly positive.

(**) $g \in \mathbf{R}[x_1, x_2, x_3]$ and for small $\epsilon > 0$, $g \geq f$ on $P((0,\epsilon))$ while $g \leq 0$ on $Q((0,\epsilon))$.

Note that (*) implies that $(0,0,0)$ is a singularity of the set of zeroes of f. If $P((0,\epsilon))$ and $Q((0,\epsilon))$ are separated by a hyperplane, then some multiple of the linear function defining the hyperplane will work for g. However, there are plenty of examples in which no such hyperplane exists. (Madden, 1988)

27. The Pythagoras number $P(K)$ of a field K is the least number n (a non-negative integer or ∞) such taht everysum of squares from K is equal to a sum of n squares. It is known that every power of 2 and every $2^n + 1$ occurs as a Pythagoras number for a suitable ordered field K. (Prestel, Jour. Reine & Angew. Math. 303/304 (1978), 284-94.) Are other natural numbers possible? (Prestel)

28. Suppose that A is a commutative ring with 1 which has no nonzero nilpotent elements, and satisfies the so-called bounded inversion property; that is, if $1 < x$ then x is a multiplicative unit. In the presence of all the other conditions, bounded inversion is equivalent to the assertion that every maximal ideal of A is convex (and therefore an ℓ-ideal.) The following conditions are then equivalent for A:

(a) If $0 \leq a \leq b$ then $a = bx$ for some $x \in A$.

(b) Every ideal of A is convex.

(c) Every ideal of A is an ℓ-ideal.

(d) The ideal generated by $a \geq 0$ and $b \geq 0$, is generated by their sum.

(e) A is a Bezout ring; that is, every ideal of A is finitely generated.

(f) Each maximal ideal M exceeds a unique minimal prime ideal $O(M)$, and each $A/O(M)$ is a valuation ring.

If A is a local ring then the above are equivalent to

(g) A is a valuation ring.

This kind of theorem has, by now, a venerable history. It was shown by Gillman & Henriksen for $A = C(X)$ in the fifties. S. Larson proves the equivalence of (a) through (d) in Can. Jour. Math., Vol. 38, No. 1 (1986), 48-64. Earlier, in de Pagter's dissertation, the same equivalences appear, for the archimedean case. There is also an example which shows that, without bounded inversion, the equivalence with (e) is lost. The theorem as it stands here is due to Martinez.

What makes the above theorem, and others like it, tick – in particular, what makes the implication (f) $\longrightarrow$ (a) work – is the assumption of bounded inversion. The space Max(A) of maximal ideals of A becomes a compact Hausdorff space (it is automatically compact, and bounded inversion makes it Hausdorff). Scott Woodward, a student of mine, has recently applied the same technique to show that if A is a Prüfer ring – if every finitely generated regular ideal is invertible – then A must be quasi-Bezout; that is, every finitely generated regular ideal is principal. (Note: a regular ideal is one which contains a non-divisor of zero.)

These arguments can also be applied to show that if every localization of A is quasi-Bezout then A too is quasi-Bezout. The converse is unknown, even in the case of $A = C(X)$. In that context this question can be reformulated as follows: if X is a quasi-F space, then is every localization of $C(X)$ quasi-Bezout? ("Localization" here is understood to mean "localization at a maximal ideal".) There is an ample supply of skepticism around to lead one to conjecture that the answer is no. (Ed. note: answer *is* no.)

29. A commutative ring with 1 is said to be <u>local-global</u> if the following property is satisfied by each polynomial over A in n variables: suppose $f(x)$ is a multiplicative unit for some substitution $x = x(M) = (x_1(M), \ldots, x_n(M)) \in (A_M)^n$,

for each localization A_M; then there is a substitution $y = (y_1, \dots, y_n) \in A^n$ so that $f(y)$ is a unit in A.

We abbreviate this property as the LG property. It has a distinguished development in commutative ring theory, from work by R.S. Pierce to that of B. McDonald and, most impressively, that of Estes & Guralnick in Jour. of Alg. 77 (1982), 138-157. It is known, for example, that if all the residue fields of A of infinite – which they certainly are for f-rings which satisfy the hypotheses of item 28 – then the LG property reduces to the following, vastly simpler property: suppose that $f(T) \in A[T]$ is a primitive polynomial (in one variable); then there is a substitution $u \in A$ so that $f(u)$ is a unit.

Consider now – as in 28 – a commutative f-ring with 1 which has no nonzero nilpotent elements and satisfies the bounded inversion property. The first condition which follows obviously implies the second:

(a) A has the LG property.

(b) For each pair $a \geq 0$, $b \geq 0$ so that a and b are relatively prime, the polynomial $f(T) = a - bT^2$ has a substitution u so that $f(u)$ is a unit.

Martinez & Woodward have recently shown that if X is a Tychonoff space, then $C(X)$ is LG precisely when X is strongly zero-dimensional space. The pleasant surprise, however, is that in this context (a) and (b) are equivalent!

For which f-rings A are conditions (a) and (b) equivalent?

30. LG rings are appealing because most of the homological properties of modules over local rings can be carried over to modules over an LG ring. Is there a similar phenomenon for f-rings (with the same hypotheses as before)? For example, can theorems about vector lattices be generalized to f-modules over LG f-rings? In

particular, it is known that an archimedean vector lattice has a unique scalar multiplication. Now if G is an "archimedean" f-module over the LG f-ring A, is the given scalar multiplication unique? (One probably need to recast the notion of archimedeaneity.) (Martinez, 1989)

31. Suppose that A is a commutative f-ring with 1, no nonzero nilpotent elements and the bounded inversion property. Suppose in addition, that 1 is a strong order unit. Then, using "Yosida" embedding techniques, one can define and f-homomorphism $\phi: A \longrightarrow C(\mathrm{Max}(A))$ the kernel of which is the Jacobson radical (it is, as well, the set of infinitesimals with respect to 1).

If A is archimedean this kernel is zero, and so A is simply embedded in $C(\mathrm{Max}(A))$. Indeed, by the Stone-Weierstrass Theorem, this is the embedding of A in its uniform completion. On the other hand, this embedding is not always as "tight" as one might think. For instance, suppose that A is a Bezout ring; then, according to the information in item 28, every maximal ideal of A contains a unique minimal prime ideal. Does the same property hold for $C(\mathrm{Max}(A))$? put differently, is $\mathrm{Max}(A)$ an F-space? The answer is no. For the ring A of eventually constant real sequences $\mathrm{Max}(A)$ is the one-point compactification of the discrete set of natural numbers. A is Bezout, but $\mathrm{Max}(A)$ is not an F-space.

The following then seem natural questions to ask:

(A) Under what conditions can one conclude from the assumption that A is Bezout? that $\mathrm{Max}(A)$ is an F-space?

(B) More generally, which kinds of embeddings of A into $C(\mathrm{Max}(A))$ preserve enough of the prime spectrum of A? (What is "enough"? Here is a reasonable candidate, from the theory of ℓ-groups: suppose that G is an ℓ-subgroup of the ℓ-group H; we say that G is a ***strongly rigid subgroup*** of H if for each $0 < h \in H$ there is a $g \geq h$ in G, so that $g^{\perp\perp} = h^{\perp\perp}$. In this setting the root system of primes of G is both an order and a topological retraction of the root system of H (an *order retraction* of a p.o. set A onto a subset B is an order preserving map π so that $\pi(x) = x$ for all $x \in B$, and if a_1 and a_2 are incomparable elements then $\pi(a_1)$ and $\pi(a_2)$ are likewise incomparable.)

Thus, if A is Bezout and A is strongly rigid in $C(\mathrm{Max}(A))$, then $\mathrm{Max}(A)$ is an F-space and conversely.

(C) For X compact, $C(X)$ has a strongly rigid subalgebra (with 1) which is hyper-archimedean if and onlyif X is basically disconnected; if so, then there is a minimal such subalgebra, namely the subalgebra of all continuous functions with finite range. (Martinez, unpublished)

Suppose we call such a strongly rigid subalgebra of $C(X)$ a *monitoring* algebra for the space X. Thus, basically disconnected spaces have (unique) minimal monitoring algebras of continuous functions. Do any other spaces have minimal monitoring algebras?

(D) Which spaces have no proper monitoring algebras? How about $C(\alpha\mathbf{N})$, where $\alpha\mathbf{N}$ is the one-point compactification of the discrete natural numbers?

To close out this array of problems, here is one which is very simple to state, but probably very difficult to solve:

32. Characterize the fields which admit a lattice order. In particular, does the field $\mathbf{C}$ of complex numbers admit a lattice order? (Schwartz, 1986)

SUBJECT INDEX

R

S

T, U

V

W

X, Y & Z

of related interest

Lattice-Ordered Groups

An Introduction

MARLOW ANDERSON

The Colorado College, Colorado Springs, Colorado, U.S.A.

and

TODD FEIL

Department of Mathematical Sciences,
Denison University, Granville, Ohio, U.S.A.

The reader will find here an up-to-date survey of the classical theory of lattice-ordered groups (l-groups).
After an initial presentation of the necessary introductory material, the four main representation theorems for l-groups are given. The following chapters explore free l-groups, varieties and torsion classes of l-groups, completions of representable and archimedean l-groups, the lateral completion, finite-valued and special-valued l-groups and groups of divisibility. A large collection of examples is included in the appendix and there is a comprehensive and up-to-date bibliography of primary references in the field. The last two sections, in particular, will be invaluable to the student and the professional mathematician alike.
The text assumes only a graduate knowledge of algebra, analysis and point-set topology and is thus accessible to graduate students and mathematicians drawn from a wide range of disciplines.

ISBN 90–277–2643–4   RTMS 4

of related interest

Lattice-Ordered Groups

Advances and Techniques

Edited by

A. M. W. GLASS

and

W. CHARLES HOLLAND

Department of Mathematics and Statistics,
Bowling Green State University, Ohio, U.S.A.

The theory of lattice-ordered groups has experienced an increasing amount of interest during the last two decades, partly because of its many applications to other areas of mathematics, such as topology, functional analysis, logic, group theory and universal algebra. The various chapters of this book deal with specialized aspects of lattice-ordered groups.
These collectively present the main results of the relevant theory and the techniques used to establish them, with emphasis on applications to other areas of mathematics. Special care has been taken to ensure that the book is readable by the specialist and by the general mathematician. An extensive bibliography is included.

MAIA 48 ISBN 0–7923–0116–1